SOTHIS PRESS

Sothis Press aims to completely reimagine the public domain book with an emphasis on universal access to captivatingly designed print editions. We believe that technology will enable every single book ever published to be perpetually accessible in beautiful print form. Sothis is beauty in ubiquity.

www.sothis.press
1 Waterhouse Square, London, ECIN 2ST

Practical Anatomy of the Rabbit

An Elementary Laboratory Textbook
in Mammalian Anatomy

B. A. Bensley

SOTHIS PRESS

ISBN 978-1-0365-1678-9

Practical Anatomy of the Rabbit
is in the public domain.

This edition is published by

The company
is registered in England and Wales.
The Company Number is 14900389.

INTRODUCTION.

As a laboratory exercise the anatomical study of an animal is chiefly a matter of applying a certain practical method of exposition, the student's attention being concentrated on those facts which can be made out by direct observation. This method is educative in the technical sense because it involves accurate discernment of detail, and, as a means of obtaining first-hand information, it is the foundation of laboratory practice.

In studying the structure of any organism, however, it is to be considered that the final object is not simply to determine in what its structure consists, i.e., its anatomy in a restricted sense, but also to understand what structure signifies, either as functional mechanism, or as the product of racial or evolutionary factors. While it is conceivable that a single organism, either as individual or species, may be considered by itself, a very superficial study suffices to show that the structure and function of no living organism can be interpreted apart from the general arrangements of organized nature, and more especially from the corresponding features of those organisms most nearly allied in point of resemblance.

This being the case, it becomes a more or less practical question in comparative study combined with dissection, or other form of laboratory practice, what the proper procedure should be. So far as the present book is concerned, it is expected that the study of the type will begin with the examination of the prepared skeleton (part II). This will be followed by dissection (part III), in which the order by sections will be found to be of less importance than that of details in any particular region. The general matter of part I is purely accessory, and though necessarily incomplete in many ways, is designed to afford a comprehensive view of the various factors upon which mammalian structure depends.

Regional sections of the foetus as figured in part I, or frozen sections of the adult animal, are a useful adjunct, since they can be used either for points of general organization, or, being substantially correct for two dimensions, can be used to remove some erroneous impressions of the position of organs incidental to their displacement in dissection.

PART I.

A GENERAL CONSIDERATION OF THE STRUCTURE OF THE RABBIT.

DIVISIONS AND METHODS.

Biology, the science or study of living organisms, includes several related sciences, the chief of which are **Anatomy**, the study of organized structure; **Physiology**, the study of function; and **Embryology**, the study of development. Anatomy, or **Comparative Anatomy**, the latter referring to the comparative study of organisms, and Embryology are also considered either as divisions, or as practical methods, of **Morphology**, the general science of the evolution of form.

The term "Anatomy" was originally applied to the dissection or study of the human body, and is still considered as referring more especially to the latter. Even in the early stages of biological science, however, the use of the term was extended to organisms generally; and afterwards, chiefly as a result of the introduction of the microscope as a new method of examining structure, it attained its present comprehensiveness as a term applying to the study of structure generally.

It has been found convenient, especially in human anatomy, to distinguish as **Gross Anatomy**, the study of that kind of structure which is displayed by dissection, or is revealed by naked-eye appearances, and as **Microscopic Anatomy**, the study of finer structure through the application of the microscope; or, again, to distinguish as **Special** or **Descriptive Anatomy**, the study of the particular features of the organs of the body, and as **General Anatomy**, the study of its more fundamental composition. General Anatomy is practically equivalent to **Histology**, the latter considering the body from the point of view of the structure and arrangement of its cells and tissues.

These distinctions are of interest in the present case chiefly as defining more exactly the practical method and the kind of structure

to be considered. Thus, **dissection** is to be recognized as a method of displaying structure of a gross and special kind. It consists in the orderly exposure and displacement of organs with the object of observing their features and relations to surrounding parts. The plan is essentially one of analysis, since conceptions of structure are based on the recognition of differences, the latter being estimated by various features, such as form, color, texture, or position. On the other hand, because of the class of structure with which it deals, dissection is also to be recognized as a preliminary method in comparison with various others involving the use of the microscope.

THE INTERPRETATION OF STRUCTURE.

Gross structure is, in a sense, only the outward expression of the finer microscopic structure underlying it, the latter being the true basis of the body. This refers not so much to the individual features of the organs as to the relation existing between their appearance as gross objects and their tissue composition. Since this relation is more fully discussed below under the head of general anatomy, it need only be mentioned here as an element in the interpretation of structure as viewed from the gross standpoint. All animal structure, however, may be considered from two points of view—physiological and morphological.

The **physiological** aspect of structure concerns the functions or activities of the living organism and of its individual parts. The contraction of a skeletal muscle is a change in the axial relations of living protoplasm, but the form and connections of the muscle are such that the contraction results in movement of one bone upon another. The excretion of urine on the part of the kidneys is the final stage of a process which rids the body of soluble waste nitrogenous materials by discharging them into a system of tubes connected with the outside of the body. What is important in these, as in a multitude of analogous cases, is that structure and function are intimately related, and in point of interpretation, serve to explain one another.

The **morphological** aspect of structure concerns various features of form and arrangement which, although they have been developed on a basis of utility, cannot be explained directly on that

basis, because the factors controlling them lie outside of the body of the individual, and are such as have operated only through a long series of gradually changing conditions in the evolution of its tpye As applied to a particular animal, the morphological method consists in explaining its adult structure by reference either to its embryonic development or to the equivalent conditions in lower existing, or perhaps fossil, forms. A recognized principle of embryology is that known as the **Law of Recapitulation** It is based on the general observation that the definitive structure of an organism is attained through a series of embryonic stages, in which it not only develops from a simple or ground type to a more complex condition but also reflects in passing the features of lower, and presumably its own ancestral, forms. That of comparative anatomy depends on the comparison of higher, specialized animals with lower, or generalized ones, the latter being assumed. in one feature or another, to have remained in a backward or primitive state of specialization, and therefore to reflect in such features a low grade of structure of a kind possessed by the ancestors of existing higher forms. These relations form a basis for the comparison of the embryonic development of organisms with the evolution or history of the groups which they represent, the former being distinguished as **ontogeny,** the latter as **phylogeny.** The interpretation of the adult structure of an organism is a matter of distinguishing its more general features from its more special ones, the former being in all cases those to which the ontogenetic and phylogenetic principles are especially applicable.

How such conditions affecting the present form of an organism have come about, may be explained by reference to ancestry. The sum of characters, apart from influences of accident, are the result of development of the primordial cell which constitutes the fertilized egg. Such features as are impressed upon the animal during growth or maturity are in this respect negligible, and the importance of the egg-cell is in no way diminished by the fact that in the majority of mammals it undergoes its early development within the maternal body. The succession of generations, or **continuity of life,** carries onward the structure of the body, and as fossil organisms reveal, has maintained this process for countless millions of years.

With succession has also come modification, the evidence of which lies not only, geologically speaking, in the relative times of appearance of life forms on the earth, but also in the fact that succession leads from primitive to specialized animals. revealing in a large way the same kinds of differences observable among those living at the present day. That the entire skeleton of a mammal is patterned upon the primitive skeleton of the fossil amphibia of the Carboniferous and Permian is evident from a comparison of the components part for part, but it is equally evident from comparative anatomy that the viviparous condition of a higher mammal is founded upon an oviparous condition in lower forms, even if no fossil evidence is forthcoming. That a mammal as an air-breathing vertebrate should develop gill structures in the embryonic condition, though circumstances never come about by which such structures are used, is in itself an important fact bearing on adult structure, but such a condition also shows to what extent a living animal carries ancestral features, whether functionally modified or not.

All characters of animals have thus an evolutionary basis, the general nature of which is easily understood although the process by which they have been developed is still a matter of uncertainty. In comparison with one another, animals present certain resemblances and differences—diagnostic features, which are used as a basis for classifying them into major and minor groups. In many cases characters of resemblance have been shown to be secondary, and are hence described as **convergent.** In some of these the resemblances are of a gross type, and the structures are described as **analogous**; in other cases they are exact or **homoplastic.** As a rule, however, characters of resemblance are broad marks of **affinity,** comparable to those seen on a small scale in human families, or in human races, and determined as in the latter cases by **heredity.** The chief basis of comparison of animals with one another is the general assumption that structures which are similar or identical are **homogenous**—of common origin On the other hand, their differences are chiefly marks of **divergence** in evolution. Although it is conceivable that many of the internal features of animals are the result of a general progressive development, more

conspicuous in comparison with those of primitive types, the majority of their differences are such as have resulted from adaptive modifications of structure, by which they have become differently adjusted to the particular conditions of their accepted habitats. **Adaptation** is one great factor in modifying animal form, producing first divergences, as between one type and its contemporaries; although such features may afterwards become settled in particular groups, and thus appear for these as **primitive, general,** or **group-characters.** Adaptation, in other words, is not a matter of present conditions only, of fixed environment, or an environment of a general or special kind. The rabbit as a gnawing animal or rodent, for example, is also an air-breathing, walking vertebrate, and shares these larger and also more ancient features with many other vertebrates of otherwise different kinds.

It is customary to include under the term **specialization** all those features in which an organism may be shown to be more highly modified in comparison with another type. If the latter is an ancestral type, or a lower form exhibiting ancestral features, its more primitive features are said to be **prototypal**, because they indicate the form from which the higher modification has been derived. Such comparisons not only reveal the fact that different animals are specialized in different degrees, but also show that a given form may be greatly specialized in some respects and primitive in others.

Moreover, it is to be considered that animals are at the present time, as they have been in the past, more or less changeable, or **plastic** types. Some of the most interesting features which they exhibit depend on the circumstance that the adjustment of structure which is rendered necessary by the opposing effects of heredity and specialization is not exact or immediate. Thus, it is not difficult to find in any specialized animal, in addition to those organs which are **functional** or in full development, others which are **retrogressive** in character and reduced in size. It is also to be assumed although difficult of proof among living forms, that there are also organs which are **sub-functional** or **progressive.**

ZOOLOGICAL POSITION.

It will be evident from the foregoing statement that every specialized animal possesses in its organization a vast assemblage of features which, if referred to their proper categories, are found to represent many grades of morphological value. In so far as the adult structure of a particular form is concerned, it is possible to consider them anatomically without discrimination; but, on the other hand, if they are to be explained, it is necessary to proceed on a basis of function, embryonic development, or evolution. The study of an animal as a **type** or representative of a group, however, concerns only in a general way the features common to its various members, since the majority of features present in any animal are of minor importance, and as such are significant chiefly as indicating the developments which may take place inside the group. The question of what an animal is actually representative is a matter of comparison with other forms, in other words, of its **zoological position.** This is expressed through the medium of **classification,** the latter being arranged to indicate, so far as is possible, the relationships of organisms to one another. In this connection the following statement of the zoological position of the rabbit may be found useful; and it may also be considered as illustrating, through the comparison of this animal with allied forms, some of the more general characters of animals as outlined above.

The domestic rabbit is represented by several races, of which the common variously-colored forms, long-haired Angoras, Lop-Ear Rabbits, and Belgian "Hares" are more familiar. They are all descendants of the wild rabbit (*Oryctolagus cuniculus*, Lepus cuniculus) of Europe. The latter is thought to have belonged originally to the countries bordering the western portion of the Mediterranean, but its distribution has been greatly extended northward and to other continents through human agency.

The family **Leporidae** contains a large number of closely related species formerly included in the single genus *Lepus*. They are variously known as hares and rabbits, but the latter designation is considered to apply more exactly to the European rabbit and its domesticated races, the others, with one or two exceptions, being more properly described as hares. The more familiar species

include the North American Cotton-Tail (*Sylvilagus floridanus*, Lepus sylvaticus), and the Prairie Hare or Jack-Rabbit (*Lepus campestris*); the European Common Rabbit (*Oryctolagus cuniculus*), and Hare (*Lepus europæus*). The two European species differ in several well marked features, which form the basis of the accepted distinctions between hares and rabbits. The rabbit is distinguished by its shorter ears and less elongated hind limbs; also by its burrowing habits, and by the circumstance that the young are born in a blind and naked condition. The hare is more nearly a running or cursorial type, and is distinguished by its longer ears—which, moreover, are tipped with black—longer hind limbs and prominent eyes. Unlike the rabbit it does not burrow, but inhabits only an open "form," and the young when born are clothed with hair and able to see.

The various species constituting this family are distinguished by several features, including the imperfect development of the clavicle, longer ears and limbs, and the presence of a distinct although greatly reduced tail, from the Picas or Tailless Hares (*Ochodontidæ*) of the mountainous districts of Central Asia and of North America (Rocky Mountains). The two families are allied, however, in the possession of a common feature, namely, the presence in the upper jaw of a second pair of incisor teeth. This feature distinguishes the sub-order **Duplicidentata** from that of the **Simplicidentata**, the latter containing the majority of rodents and embracing all forms with a single pair of upper incisors.

The mammalian order **Rodentia**, to which the family belongs, contains a very large assemblage of forms—the Squirrels, Marmots, Cavies, Beavers, Mice, and Porcupines being among the more familiar. This order is distinguished by the modification of the anterior incisors in both upper and lower jaws to form chisel-like cutting organs, the teeth having their enamel layer disposed chiefly if not wholly on their front surfaces, so that they remain in a permanently sharp condition. This modification is associated with an extensive obliteration of intermediate teeth, comprising posterior incisors, canines, and anterior premolars; also with elaboration and often great complexity of the remaining premolar and molar teeth, the lower jaw, and, indeed, the parts of the skull generally. Characteristic of these animals is the extension, both forward and

backward, of the jaw-musculature. The articulation of the lower jaw exhibits an elongated articular process fitting into a corresponding longitudinal fossa on the skull, the jaw being able to move forward and backward in addition to vertically and from side to side. The teeth are further arcuate in shape, and are provided with open roots, so that their growth is not limited, as it is in the majority of mammals. The rodents are in many particulars primitive types. For example, they tend to retain the five-toed (pentadactyl), plantigrade foot, characteristic of primitive mammalia and, indeed, of terrestrial vertebrates, and exhibit also unelaborated cerebral hemispheres in the brain. In other respects, however, as in the rodent characters above-mentioned and in the elaboration of the intestine, especially the caecum, they exhibit the characters of highly specialized herbivores.

Like all higher or **placental mammalia** (**Placentalia**), the rabbit is viviparous, the young being retained through a period of gestation in the maternal uterus, to the wall of which they are attached by a vascular connection, the placenta. In this feature the placental mammalia differ from the **marsupial mammalia** (**Marsupialia**) of Australia and South America, the latter being viviparous, but, with one exception, without placenta; also from the **egg-laying mammalia** (**Monotremata**) of Australia, the latter being oviparous, like the majority of the lower, reptilian forms. These three sub-classes of mammals are united, however, by the common features of the class **Mammalia.** They are warm-blooded animals, provided with a complete double circulation, and with a hairy investment for the surface of the body. In all, the young are nourished for a time after birth through the secretion of modified cutaneous, milk-producing, or mammary glands.

Many of the more general features of the rabbit are such as are not recognized by group designation, but yet are shared with other terrestrial vertebrates, including mammals, reptiles, birds, and, in part, amphibians. This refers to the development of the lungs and associated respiratory tracts, both the true respiratory tracts and the accessory respiratory passages traversing the skull; further the loss of the branchial or fish-type of respiration and the new disposition of the branchial structures; the development of a tri-

segmented type of limb with a full complement of muscles, and originally a pentadactyl, plantigrade foot, for support of the body and for locomotion. The regional differentiation of the vertebral column, especially the mobility of the neck, the free occipital articulation, and the definition of the sacrum, the latter associated with the elaboration of the pelvic girdle, are all features of general signifiicance in the terrestrial vertebrates.

Finally, the rabbit agrees with other members of the phylum **Chordata** in the possession of a ground-plan underlying the most general features of its organs, and the position, arrangement, and plan of development of its organ-systems. All **Vertebrata** or back-boned animals possess an axial skeleton formed by the segmented vertebral column. In a very comprehensive way they possess as chordates a still more fundamental axial support, the notochord, the latter being an embryonic structure except in the lowest chordates. In a more restricted sense, as **Craniota**, they possess an organized head region with differentiated brain, special sense organs, and enclosing primary skull. They possess a series of branchial (branchiomeric) structures, appearing either in the adult condition, as in fishes, or as part of the underlying plan in the embryonic condition; and they add to their general features in the arrangement of the organ-systems the further feature of transverse segmentation (metamerism) of a considerable portion of the body.

These facts may be set down in tabular form, as indicated below. A similar plan can be constructed for any group of organisms, but whether it constitutes a natural or an artificial classification depends on whether or not it is based upon an actual study of the affinities of the organisms concerned. A natural classification should show at a glance not only what the relative importance of any particular character may be, but also how it stands in the scale of specialization. For example the placental stage of vertebrate development, *i.e.* the development of the placenta itself in the highest stage of vertebrate evolution, is the culmination of a series of arrangements for the care of eggs and young, and further the adherence of human structure in the vast majority of features to type of higher mammals is similarly expressed by the fact that man is also a placental mammal.

Phylum **CHORDATA.** Animals with notochord and gills.

† PROTOZOA, ANNULATA, MOLLUSCA, etc., invertebrate phyla.

Sub-Phylum **CRANIOTA.** Chordates with organized head region.

† ACRANIA. Lancelets. Also Ascidians and worm-like Chordates, sometimes separately classified.

VERTEBRATA.

Class **MAMMALIA.** Warm-blooded craniotes, with hair coat. Young nourished from mammary glands.

† PISCES, AMPHIBIA, REPTILIA, AVES, lower vertebrate classes.

Sub-Class **PLACENTALIA.** Mammals with placenta.

† MARSUPIALIA. Viviparous mammals without placenta.

† MONOTREMATA. Oviparous mammals.

Order **RODENTIA.** Gnawing placentals, with chisel-like incisors.

† CARNIVORA, UNGULATA, PRIMATES, etc. Various placental orders otherwise adapted.

Sub-Order **DUPLICIDENTATA.** Rodents with two pairs of upper incisors.

† SIMPLICIDENTATA. One pair of upper incissors. Squirrels, Gophers, etc..

Family **LEPORIDAE.** Hares and Rabbits.

† OCHODONTIDAE. Picas).

†*Equivalent groups.*

GENERAL ANATOMY.

Although in every respect a continuous structure, the body is differentiated into a large number of parts, or **organs**, the latter being more or less individual in form, composition, or function. Organs are arranged for the most part in **systems**, each of which is concerned with some general or fundamental function, to which several organs may contribute.

In a more general way the body may be considered as an assemblage of **tissues**. The latter are layers or aggregations of similarly differentiated cells. They are of several different kinds and are variously associated in the formation of organs. Being structures of an intermediate position they may be considered either as organ components or as products of specialized cells.

As a body-unit a **cell** consists of a small mass of living **protoplasm**, containing a central body, the **nucleus**, and surrounded or enclosed on its free border by a **cell-membrane**. The nucleus is a highly organized body, having an important function in the reproduction of the cell and also in its general activity or metabolism. It contains a characteristic formed material, **chromatin**, and frequently also a minute spherical body, the **nucleolus**. The chief features of a typical cell are illustrated in the accompanying figure (1) of the developing ovum, the latter being a single cell, noteworthy for its large size, and also one in which the external form is not greatly modified, as it is in the majority of the cells of the body. Its enclosing membrane, the zona pellucida, by which in its natural position in the ovary it is separated from the surrounding follicular cells, is considered to belong in part to the latter.

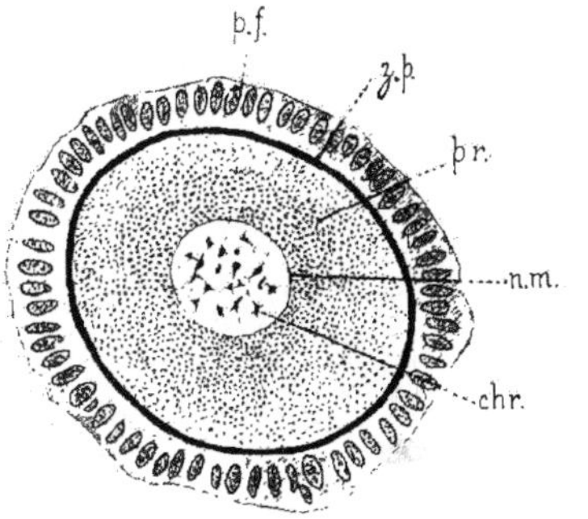

FIG. 1. Developing ovum of the rabbit. From a section of the ovary: chr., chromatin; n.m., nuclear membrane; p.f., cells of the primary ovarian follicle; z.p., zona pellucida; pr., protoplasm.

As fundamental living matter, protoplasm possesses certain

Considered collectively, these functions are not so well illustrated in the higher or **multicellular** organisms, in which particular functions are assigned to particular cells, as in the lower **unicellular** organisms, in which all functions are discharged by a single cell. In simple or protozoan animals the protoplasm is seen to be capable of ingesting food-materials, of discharging waste, of changing its form, and of reacting in one way or another to stimuli arising outside of the body. Moreover, the protozoan cell is capable of giving rise to new cells by division of its substance into two parts, which process originates in the nucleus, and is associated at some stage usually at least, with union or **conjugation** of parent cells.

All the cells of the body of a multicellular organism are products of a single cell, the **fertilized egg,** but the latter is a product of fusion of two primary elements, the **spermatozoon** of the male parent and the **ovum** of the female. The fertilized egg does not exhibit the functions of a one-celled body, but possesses the potential of these functions, and the latter appear, to a large extent individually, in the differentiation of its division-products into specialized tissue-elements.

It is in this way that the body of a multicellular animal must be founded with reference to the same elementary functions of life as those appearing in one-celled organisms. But the repeated division of the fertilized egg, in development toward the adult condition, gives rise by **division of labor** to a great variety of cells, each kind of which may be regarded as representing a minor aspect of some major function.

The Tissues.

The primary tissues of the body are of four kinds—**epithelial, connective, muscular,** and **nervous.** To these—the **fixed** tissues—are to be added the fluid substances, **blood** and **lymph,** in which the cell elements, the red and white corpuscles, or in the latter case the white elements alone, are suspended in a fluid medium.

The following survey of the principal features of the tissues will serve to make clear the extent to which the gross appearance of organs depends upon tissue composition, though the account itself is in no way intended as a guide to the microscopic structure of the rabbit, which is more properly part of the subject matter of histology.

Epithelial Tissues.

Epithelial tissues are distinguished chiefly as surface investments, such as those of the exterior of the body, the interior of the alimentary canal, the lungs, the respiratory and accessory respiratory tracts, and the ducts of the urinogenital organs. In all epithelia the **cellular** feature is a prominent one, and it is largely for this reason that as lining membranes they are not conspicuous in gross structure. They are noteworthy, however, for their products, the hairs and the various kinds of secreting organs or **glands.**

The epithelium of the skin (Fig. 2) is known as the **epidermis** or scarf-skin. It is **stratified,** i.e., composed of several layers deep of cells, of which the deeper are **formative,** growing cells, while those at the surface are flattened **squamous** cells, and are successively discarded. The several layers combined produce but a thin membrane. It extends over the entire surface of the body and connects at certain points with the epithelia of the internal surfaces. It is supported by a thick resistant layer of connective tissue which forms the true skin or **corium.**

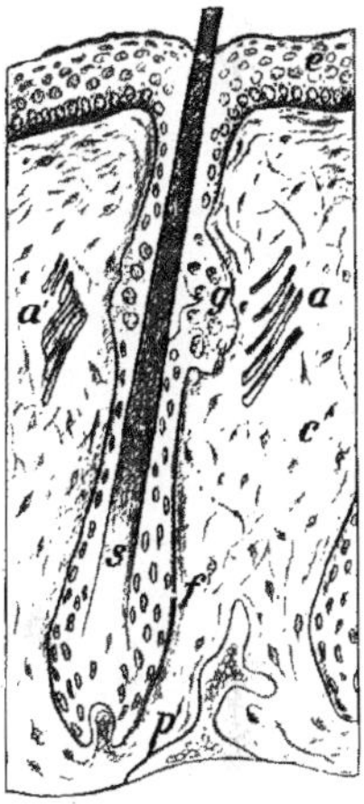

FIG. 2. From a section of the upper lip of a four-day-old rabbit. a,a', erector muscles of the hair; c, connective tissue of corium; e, epidermis; f, hair follicle; g, sebaceous gland; p, papilla; s, hair shaft.

The epithelium of the internal surfaces forms the chief portion of the **mucous membranes.** In the greater portion of the alimentary tract the epithelial layer is **simple** or one layered and is associated with a thin layer of smooth muscle to form a **mucous tunic** (Fig. 16, t.ms.).

The coating of hairs on the surface of the body, the presence of which is a notable mammalian feature, is a protective investment arising from the epidermis. A **hair** is produced by the modification of the central portion of an ingrowth of the epidermis, termed the **hair follicle** (Fig. 2, f). The latter contains at its base a small elevation of the underlying vascular connective tissue, the **hair papilla,** through which the structure is nourished. On

the general surface of the body the hair follicles are arranged in groups (Fig. 5), and on the lips certain large follicles are set apart for the production of the greatly enlarged sensory hairs or **vibrissae.** Connected with the hair follicles are thin strands of smooth muscle, the **arrectores pilorum** (Fig. 2, a,a′). They are placed in the broad angles formed by the inclined follicles with the corium and their contraction throws the hair into a more nearly erect position.

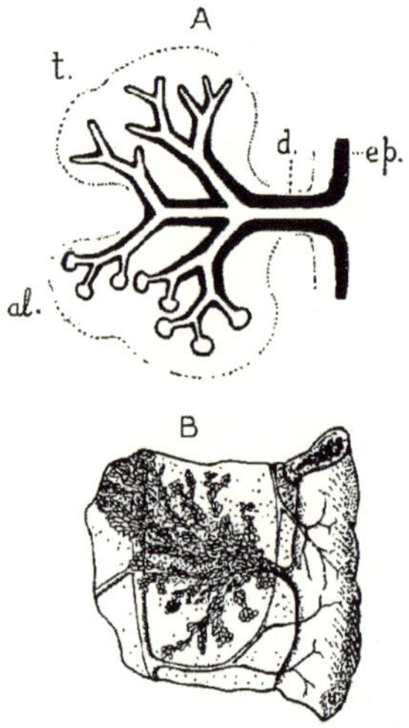

FIG. 3. A, Diagram of a multicellular gland: al., alveolar type; d., duct; ep., general epithelium; t., tubular type. B, The pancreatic duct and associated portion of the duodenum in the rabbit.

Epithelial glands are ingrowths of the general layer, the cells of which become greatly modified as **secreting** structures. The **lumen** or cavity of the gland, in most cases greatly complicated through the division of the gland substance, is connected with the general surface by a **duct** which serves to carry away its secretion. In some cases the connection of a gland with the epithelial surface is embryonic, and in the adult condition the gland is found separated from the epithelium from which it was originally formed. This condition is represented by the thyreoid and thymus glands of the neck and thorax respectively. In other cases the secreting element is a single cell, the latter thus representing a **unicellular** gland lying directly in the general layer. The mucus-secreting goblet cells of the intestinal wall are structures of this nature. Typical epithelial glands are accordingly **external secreting** glands in the sense that the secretion is discharged and utilized on a free surface. There are many structures, such as the suprarenal and pituitary bodies, commonly described as **internal** secreting glands, the products of which as **hormones** or chemical bodies of various kinds have a modifying effect upon functions in outlying parts of the body. They are not of necessity epithelial in origin and their resemblance to ordinary glands is usually only a matter of superficial appearance.

The majority of **multicellular glands** conform to one of two types, namely, the **tubular** gland, in which the secreting portions are of uniform calibre, and the **acinous** or **alveolar** gland, in which the secreting portions are sacculated (Fig. 3, A). Both types occur in simple, little branched and greatly branched conditions (Figs. 3, B, 4).

Cutaneous glands of two types are commonly present in mammals in association with the hairs, namely, **sudoriferous** or sweat-glands, which are glands of the tubular type, and **sebaceous** glands, which are of the acinous type. In the rabbit, glands are absent from the general surface, but are found in special situations, as, for example, in connection with the hair follicles of the lips, the internal surface of the ear, and the external genital organs. The **inguinal glands** comprise both tubular and acinous portions. The **mammary glands of** the female are greatly modified cutaneous glands of an acinous type.

Apart from the mucus-secreting cells of the general epithelium, the glands of the alimentary canal comprise the important but less elaborated glands of the wall, such as the **gastric glands** of the wall of the stomach; and the greatly elaborated, outstanding glands which lie beyond the wall and are connected with the interior of the canal only through their ducts. The latter comprise the **oral glands,** the **liver,** and the **pancreas.** The oral glands include chiefly the **submaxillary, parotid, sublingual,** and **infraorbital glands**—conspicuous structures in the dissection of the surrounding portions of the head and neck. Their ducts communicate with the cavity of the mouth.

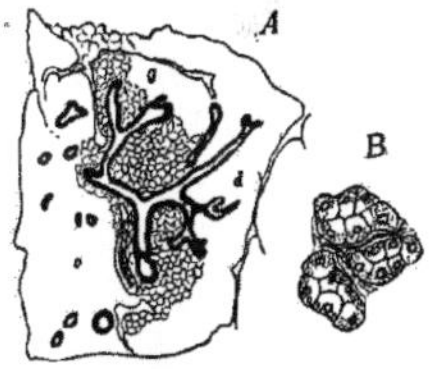

FIG. 4. From a section of the parotid salivary gland of the rabbit. A. Duct system in relation to body of gland: g, acini of gland body; d, interlobular duct; B. Three individual acini, highly magnified.

The secretion of the oral glands has important mechanical functions in moistening the comminuted food in preparation for swallowing. It also contains ferments or **enzymes,** chiefly ptyalin, which is capable of transforming starch in soluble materials, though the digestive action is probably not exercised to a great extent. Pancreatic secretion, on the other hand, has little or no mechanical action, but its enzymes are of the greatest importance in digestion.

There is a close association between the epithelia of the surface of the body and the nervous tissues, arising from the circumstance that they are derivatives of a common embryonic layer, the ectoderm. In the adult we may distinguish as **sensory epithelia** special aggregations of cells lying either in a deep or superficial position, and associated more or less closely with the central nervous system. They comprise the **olfactory epithelium** of the nasal cavity, some of the cells of which are true nerve cells, the **gustatory epithelium** of the tongue, and the **auditory epithelium** of the membranous labyrinth of the ear. The **retina**—the nervous portion of the eye—is a modified portion of the central nervous system.

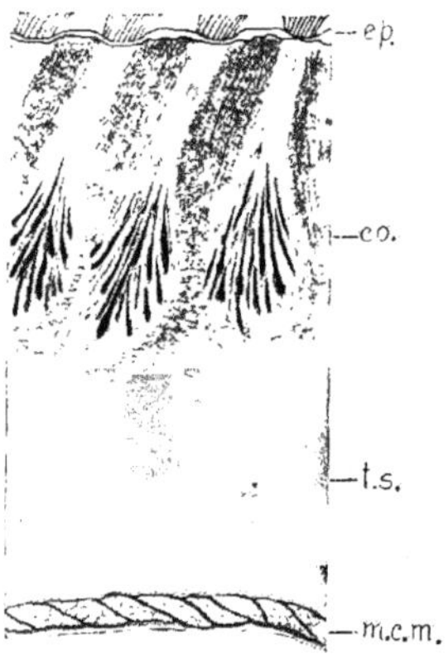

FIG. 5. From a section of the skin of the side of the body of an adult rabbit. × about 10; showing the grouping of the hair-follicles: co., corium; ep., epithelium; m.c.m., cutaneus maximus muscle; t.s., subcutaneous tissue.

As linings of surfaces, the ordinary epithelia may be distinguished from certain special coverings of internal spaces, the **endothelia** and **mesothelia**. The latter consist microscopically of thin pavement-like cells. They differ from epithelia in origin, being formed, not in connection with originally free surfaces, but in relation to spaces of the mesoderm or intermediate layer of the body. Endothelia form the linings of blood vessels and lymph canals, while mesothelia are the chief layers of the smooth, moist **serous membranes** which line the peritoneal, pleural, and pericardial cavities.

2. Connective Tissues.

The **connective tissues** form the supporting elements of the body. As **ordinary connective** tissues they serve to connect organs or parts of organs together, and as **skeletal** tissues they provide the rigid framework or skeleton from which all soft parts of the body are suspended. They are distinguished by the presence of two main components—the **cell basis**, and the **intercellular**

substance or **matrix.** The cellular portion is formative, and is much more conspicuous in the embryonic than in the adult condition. All connective tissues are products of an embryonic tissue, the **mesenchyme** (Fig. 23, ms.), which consists of branched cells connected together by their outstanding processes. Through the activity of the cells there is formed an intercellular material consisting either of a homogeneous matrix, or more frequently a matrix containing formed elements of a supportive nature.

A. Ordinary Connective Tissues.

In the adult condition the ordinary connective tissues, with few exceptions, consist of the cell basis with two kinds of fibrous elements, the **white,** and **yellow fibres** (Fig. 6). White fibres are elements of great strength. They are fine, unbranched fibres, which do not occur independently, but are joined in a parallel fashion with one another, forming in this way fibre bundles of larger or smaller size. The yellow fibres are of greater diameter. They branch and communicate, but are not associated to form bundles. They also differ from white fibres in being highly elastic. The tissue produced in this way is known as **fibrous connective tissue.** It occurs in several forms according to the relative concentration of the two kinds of fibres or the admixture of other materials.

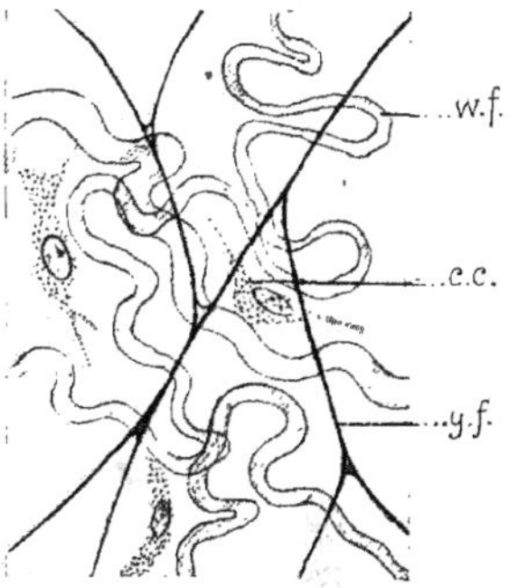

FIG. 6. Areolar connective tissue (subcutaneous tissue) of the rabbit; from an embalmed specimen: c.c., connective tissue cell; w.f., bundle of white fibres; y.f., yellow elastic fibre.

The commonest kind of fibrous tissue in the adult is that described as **areolar.** It is characteristic of the **subcutaneous tissue** (Fig. 5) which connects the skin with the body; but areolar tissue occurs also in various positions where it has a similar function of joining structures loosely together. Subcutaneous connective tissue is a white material, the peculiar appearance and properties of which are due to the fact that the two kinds of fibrous elements are arranged in a loose felt-like network (Fig. 6). When stretched

it is found to yield up to a certain point, beyond which it is tough and resistant. It tends to regain its original shape when the tension is removed.

Fibrous connective tissue may be greatly modified through the concentration of either one of the fibrous elements. Concentration of white fibres is most common. This condition is illustrated in the thick connective tissue layer forming the true skin or **corium** (Fig. 5), but is more conspicuous in the glistening white **tendons** (Figs. 7, 34) by which muscles are attached to bone surfaces, in the **ligaments** uniting bones with one another, and in the thin, broad **aponeuroses** which serve for muscular attachment. The structures known anatomically as **fasciae** are special sheets of connective tissue covering chiefly individual muscles or muscle groups. Concentration of yellow fibres occurs in the dorsal ligament of the neck (ligamentum muchae) where greater elasticity is required. The ligament is not so conspicuous in the rabbit as in larger mammals where the yellow coloration is very noticeable.

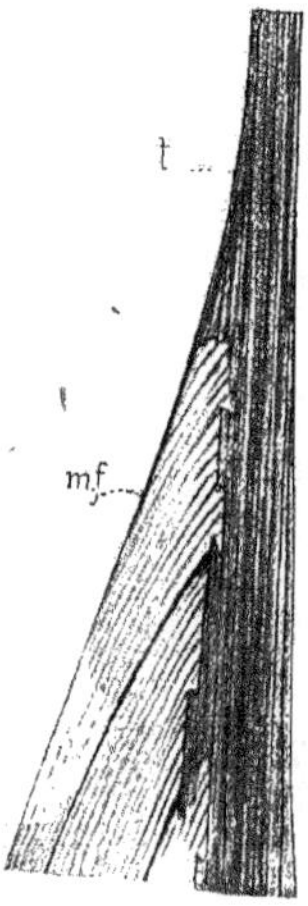

FIG. 7. From a section of the tendon of origin of the biceps muscle: m.f., muscle fibres ending on the tendon; t., fibrous connective tissue of the tendon.

Fat or **adipose tissue** is a soft form of connective tissue in which the cells are greatly enlarged and contain fatty material in the form of globules. It tends to occur in certain situations and in particular association with the bloodvessels, but otherwise is found in situations where areolar connective tissue might be expected to occur. Special fat masses, sometimes distinguishable by their darker coloration, occur at the side of the neck in the rabbit. In the foetus (cf. Plate VI) large masses of vascular connective tissue are found in this situation, and are probably similar in origin to the storing or hibernating glands of certain other mammals.

Coloration or **pigmentation** of certain portions of the body, especially of the skin and hairs, and of the retina, ciliary body, and iris of the eye, is due partly to the presence of special connective

ssue cells, **chromatophores,** and partly to the presence of gment granules in epithelium. The absence of pigmentation in iimals belonging to species normally colored—**albinism**—is dicated by the white coloration of the hair and by the pink ›loration of the eyes, the tter being due to the cir-ımstance that the blood-essels of the vascular tunic e not concealed by pigment.

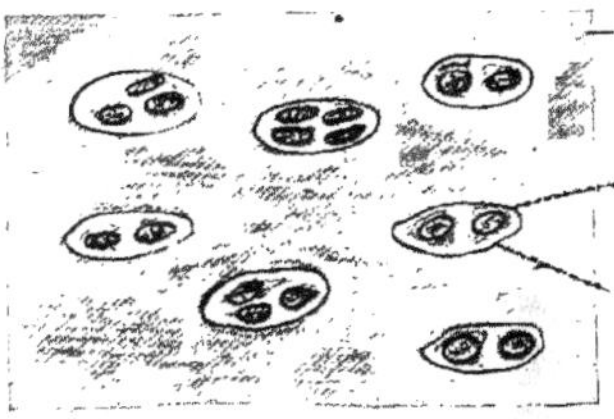

FIG. 8. Hyaline cartilage: c.c., cartilage cell; l., lacuna; m., matrix.

B. Skeletal Tissues.

The skeletal tissues are ›lid forms of connective ssue which, on account of ıeir more permanent shape, are better adapted to form a support ›r the body. They are of two kinds—**cartilage** and **bone.**

Simple or **hyaline cartilage** (Fig. 8) is a semi-solid and some-hat resilient material of a bluish or pearly coloration. It consists a homogeneous **matrix** in which the cells are imbedded. The ›lls are distributed singly, more often in groups of vo to four, each group being ›ntained in a small oval ›ace, the cartilage **lacuna.** he size of the spaces, and so their distance apart, is ıbject to great variation. he addition of fibrous ele-ıents, usually of white fibres, roduces a modification nown as **fibrocartilage.** /hile fibro-cartilage occurs certain situations, as in the ʳmphysis of the pelvis, or in connection with the interarticular enisci and at the capsular margins of the joints.

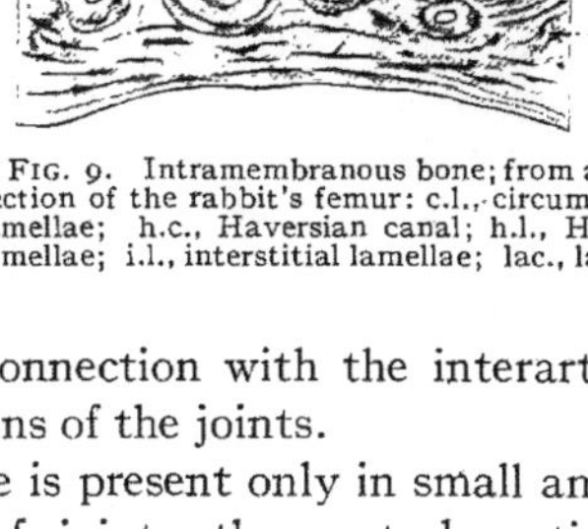

FIG. 9. Intramembranous bone; from a ground section of the rabbit's femur: c.l., circumferential lamellae; h.c., Haversian canal; h.l., Haversian lamellae; i.l., interstitial lamellae; lac., lacuna.

In the adult skeleton cartilage is present only in small amount. forms the articular surfaces of joints, the ventral portions or

:ostal cartilages of the ribs, and a portion of the nasal septum; t is also found uniting the basal bones of the skull. In the embryo, ıowever, it forms the entire skeleton, with the exception of a small ɔortion which, as described below, is formed of membrane bone. n the course of development, except in the situations indicated, he cartilage is replaced by bone.

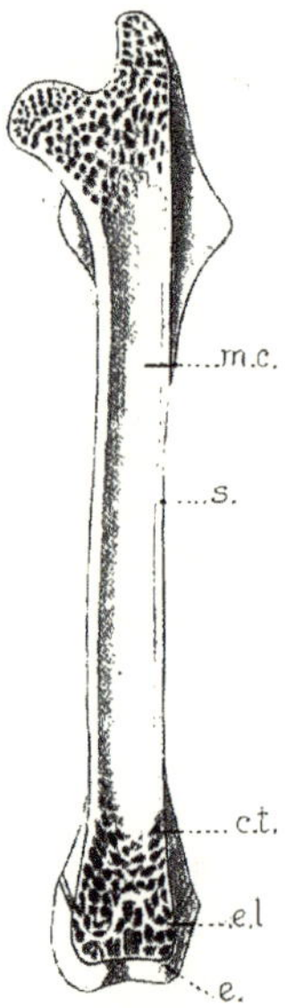

FIG. 10. Divided femur of the rabbit; c.t., cancellous tissue; e., distal epiphysis; e.l., epiphysial line; m.c., marrow-cavity; s., shaft.

Bone is a compact, resistant, but yet somewhat elastic tissue, possessing much greater strength than cartilage, and therefore forming a more perfect skeletal support. As indicated below, its appearance as a tissue differs somewhat according to its mode of formation. The more typical structure (intramembranous bone) is illustrated in the accompanying figure (9) of a ground transverse section of the dried shaft of the femur. Its dry weight consists of about one-third animal matter and two-thirds mineral matter, the latter being chiefly calcium phosphate. The bone materials are deposited in layers or **lamellae,** which are comparable to highly modified white fibres of fibrous connective tissue. The lamellae enclose between them greatly branched spaces, the **lacunae,** in which in the natural condition the **bone-cells** are accommodated. On account of certain differences in development, the lamellae are disposed in different ways. Some are disposed concentrically around longitudinal spaces, the **Haversian canals,** forming in this way the so-called **Haversian systems.** The canals are occupied in the natural condition by bloodvessels. Others of them, the **interstitial lamellae,** are disposed in a somewhat parallel fashion between the Haversian systems; while a third series, the **circumferential lamellae,** are disposed in a parallel fashion with respect to the periphery or the internal surface of the bone. In the natural condition the bone is enclosed except on its articular surfaces by a layer of connective

tissue, the **periosteum**. During the period of growth this membrane contains large numbers of bone-forming cells, the **osteoblasts**, through the activity of which the deposition of the bone lamellae is accomplished.

It is only in a few cases that the bones of the skeleton may be looked upon as solid structures—as a rule they consist of a fairly thin shell of hard or **compact** bone surrounding a central mass of spongy or **cancellous** bone. This arrangement is one of great mechanical strength, combined with lightness, and at the same time provision is made in the interior of the bone for bloodvessels and marrow spaces. Thus in a long bone (Fig. 10) the central portion or **shaft** consists of a cylinder of compact bone surrounding an extensive space, the **marrow-cavity**. The ends or **extremities** consist each of a thin shell of compact bone continuing that of the shaft and surrounding a mass of cancellous tissue. In the short, flat, or irregular bones of the skeleton no continuous marrow cavity is formed.

In point of origin bones are of two chief kinds—**membrane** or **derm bones**, and **cartilage** or **replacing bones**, but a third group is represented by the tendon or **sesamoid** bones which are developed in the tendons of muscles. The membrane bones are few in number. They comprise the roofing and facial bones of the skull, most of which are distinguished by their flattened, tabulate character. They are formed in connective tissue membranes, and although they sometimes contain cartilage they are not formed on a cartilage basis. Moreover, they are elements lying in a superficial position with respect to the skeleton proper. They owe this condition, and also their designation as derm bones, to the circumstance that they represent surface plates which in lower vertebrates are associated with the skin.

Cartilage bones, on the other hand, are the characteristic elements of the skeleton. They are known as replacing bones because they are formed on a cartilage basis, the latter, as indicated above, representing the primary cartilage skeleton of the embryo. In the course of development, except in certain regions where the cartilage persists throughout life, the cartilage material is replaced by bone, which tends to surround and also invade it.

The way in which bones are formed on a cartilage basis serves to xplain many peculiarities of the adult skeleton. In the embryonic ondition the cartilage rudiments are associated to form a com- lete but primitive skeleton. In many cases the replacement of hese elements by bone is not direct, certain readjustments being ecessary, both for purposes of growth and on account of the much ıore special functional requirements of the adult skeleton.

In the embryonic condition the cartilage rudiments are en- losed by a connective tissue sheath, equivalent to the periosteum f a bone, but described as the **perichondrium.** The osteoblasts f this layer are concerned with the formation of bone material, oth in the interior (**endochondral bone**) and on the surface **ntramembranous bone**). The formation of endochrondral bone is preceded by certain changes which take place in the interior of the cartilage. In the latter, in certaifi areas, known as the **centres of ossification,** the matrix becomes partly dissolved, the cells enlarged and ultimately broken down. These changes are associated with a deposition of calcareous material, or **calcification,** by which the portion of the cartilage undergoing transformation is temporarily strengthened. Into this area the active cells of the erichondrium are carried through the agency of vascular in- owths, the **periosteal buds,** and the result of their presence is ıe deposition of bone material in association with the remaining ortions of the matrix. This condition is partly illustrated in the istal epiphysis of the humerus shown in Fig. 13, A, the figure being om a vertical section of the elbow-joint of a four-day-old rabbit.

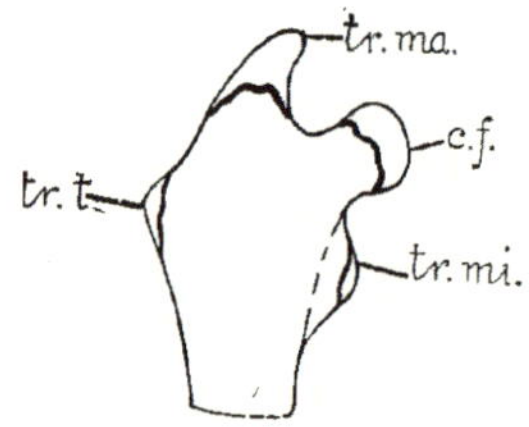

FIG. 11. Outline sketch of the prox- al end of the femur of a young imal: c.f., principal epiphysis for the ad of the femur. The accessory iphyses are for the great (tr.ma.), ser (tr.mi.), and third (tr.t.) tro- anters.

In the long bones the formation of the first or **main centres** of sification takes place in the shaft, and there are formed afterwards cessory or **epiphysial centres** for the extremities. A divided tremity, such as the proximal end of the femur (Fig. 11), may ssess several such centres—a principal one for the chief epiphysis actual extremity of the bone and several subsidiary centres for

its outstanding processes. In the shaft the formation of endochondral bone is of short duration. Through the activity of the osteoblasts lying directly in the perichondrium, or later the periosteum, a process of formation of intramembranous bone goes on during the whole period of growth, and the result of the peripheral deposition of bone lamellae is, as described above, that the transverse diameter of the bone is greatly increased. The enlargement of the marrow cavity, with which this is associated, is produced by the absorption of bone from the interior.

In young animals both the epiphysial centres and the masses of cartilage in which they are formed are sharply marked off from the body of the bone (cf. Fig. 13). This is largely because the formation of the epiphysial centres tends to lag behind that of the main centres, and thus the cartilage extremities of the bones are evident long after the formation of the shaft is under way. In the epiphysial centres the bone formation is endochondral. The bone masses which they form are distinguished as **epiphyses.** During the period of growth they are connected with the body of the bone by plates of epiphysial cartilage, into which the surrounding perichondrium extends as an ossification ridge. In this region bone formation takes place, with the result that the whole structure is greatly increased in length.

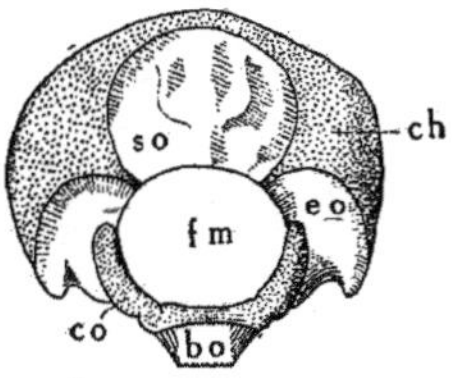

FIG. 12. The occipital portion of the skull in a three-day-old rabbit. bo, basioccipital bone; ch, occipital portion of chondrocranium; co, occipital condyle; eo, exoccipital; fm, foramen magnum; so, supraoccipital.

After the period of growth, the duration of which differs in different bones, the epiphyses become firmly co-ossified with the body of the bone, although the lines of junction or **epiphysial lines** may be still visible. Thus in the distal extremities of the radius and ulna, in the proximal extremities of the fibula, or in the bodies of the lumbar vertebrae, the epiphysial lines appear even in old animals. In the foregoing figure (10) of the divided femur it will be seen that the position of the epiphysial lines is indicated by bands of compact tissue. Finally, in thoroughly macerated bones of young animals, the epiphyses are usually found to be readily separable from the bones.

In a comparison of the adult skeleton with the more primitive embryonic skeleton, several differences in the arrangement of the elements are evident. Thus many bones, notwithstanding their possession of several centres of ossification, are to be looked upon as individual either in the cartilage or in the bone condition. In other cases, as in the basal portion of the skull, separate bone elements are produced in a mass of cartilage primarily continuous. These either remain distinct throughout life, or, as in the occipital region, (Fig. 12), become fused together to form compound or **composite bones.** In still other cases, as in the vertebrae, the apparently single elements of the adult condition are the products not only of originally distinct bones, but also of primarily separate cartilage masses.

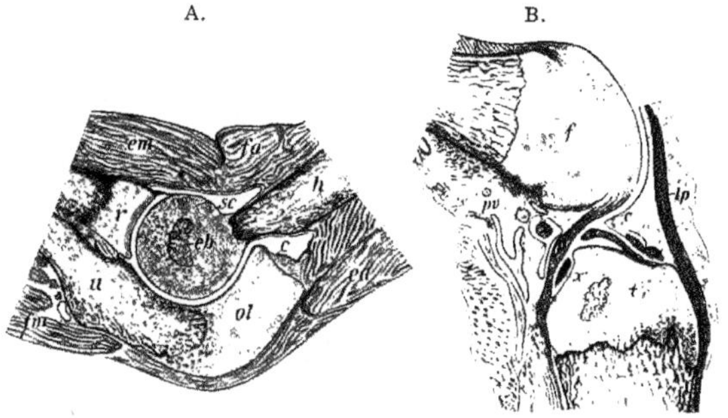

FIG. 13. Vertical sections of elbow and knee of four-day-old rabbit. A elbow; c, capsule; eb, endochondral bone in the distal epiphysis of the humerus; ea, extensor muscles of the forearm; em, extensors of the hand; fa, flexors of the forearm; fm, flexors of the hand; h, humerus; ol, olecranon; r, radius; sc, synovial cavity; u, ulna. B, knee; a, anterior cruciate ligament; c, capsule; f, femur; lp, patellar ligament; p, posterior cruciate ligament; pv, popliteal vessels; t, tibia; x,x, anterior and posterior ligaments of the lateral meniscus; x',x', anterior and posterior ligaments of the medial meniscus.

The bones of the skeleton are united or **articulated** with one another by connective tissue in the form of ligaments, by cartilage, or in some cases by both together, i.e., by fibrocartilage. Ligamentous union, distinguished as **syndesmosis**, is the most general type of articulation. Cartilage union or **synchondrosis** occurs in certain situations, as in the basal region of the skull. Union by fibrocartilage or **symphysis** is characteristic of the articulation of the two sides of the pelvis (symphysis pubis).

The articulations of bones are of two types—immovable articulations or **synarthroses,** and movable articulations, **diarthroses,** or joints. In the former, motion is either absent or at least greatly restricted. In the latter, it is definitely provided for through the presence of joint-structures. Thus in a joint (Fig. 13) the apposed surfaces of the bones are accurately modelled in relation to one another, and are moreover covered by a layer of cartilage, the latter forming a joint cushion. Between the two surfaces is a space, the cavity of the joint, containing a viscid material, the **synovia,** which serves for lubrication. The space is enclosed by a connective tissue **capsule.** The strength of the joint depends largely on the enclosing capsule, but it is usually greatly increased by the presence of accessory ligaments. In the more complex joints, such as that of the knee (Fig. 13 B), **interarticular cartilages** (menisci) are enclosed between the bone surfaces, and the latter are connected directly by short ligamentous cords. The various ligaments of a joint permit free motion of the bones, but only up to a certain point.

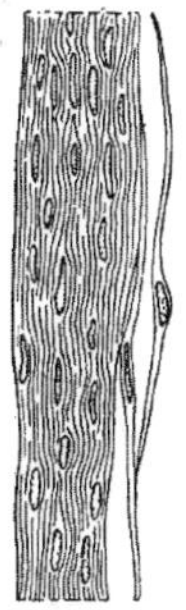

FIG. 14. Involuntary muscle, from a section of the muscular tunic of the intestine.

Several differences are observable in joints according to the form of the apposed surface and the kind of motion provided for. Thus in the ball-and-socket joint or **enarthrosis,** exemplified by those of the shoulder and hip, a bone is able to move in various directions about its base of attachment, although actually, in the limbs, this motion is almost restricted to an anteroposterior direction. In the **ginglymus** or hinge-joint, as exemplified by the distal articulations of the limb, motion is restricted to a single plane. The gliding joint or **arthrodia** is one in which a slight degree of motion is made possible by one surface slipping over the other; it is exemplified in the accessory articulations of the vertebral arches.

3. Muscular Tissues.

Muscular tissues are the active portions of the individual muscles of the skeleton and of the muscle coats of visceral organs. Their chief feature consists in the elongation of the cells to form

fibres. These fibres may be considered to possess the contractile properties of protoplasm, but with the contraction limited to one direction. Except in a few cases the fibres are arranged in a parallel fashion, so that the line of contraction of the muscle or muscle layer is the same as that of each of its fibres. The result of contraction in both is the shortening of the longitudinal axis and the increase of the transverse axis. Muscles are important structures in the production of heat. The latter is liberated not only in action but also when the muscle is in repose.

Two chief types of muscle fibres occur in the body—the **smooth** or **unstriated** fibres, which are characteristic of the **involuntary muscles** or muscle coats of the visceral organs or of the skin, and the **striated fibres** which compose the individual or **voluntary muscles** of the skeleton. Smooth fibres (Fig. 14) are elongated, spindle-like cells, the substance of which is longitudinally striated, but possesses no transverse markings. The single nucleus of the cell occupies a central position. The muscles which they form are distinguished as involuntary because their operation is not under the control of the will, their connections being with the sympathetic nervous system. The striated fibres (Fig. 15) are very much larger, cylindrical structures, the substance of which possesses characteristic transverse striations. Each fibre is enclosed by a loosely attached membrane, the sarcolemma, on the inner surface of which many nuclei occur.* The presence of these nuclei indicates that the fibre is not a single cell but a syncytium, *i.e.*, an association of cells unseparated by cell boundaries. The muscles formed by such fibres are under the control of the will, their connections being directly with the central nervous system. They comprise not only the typical muscles of the skeleton, but also the special muscles connecting the skeleton with the skin.

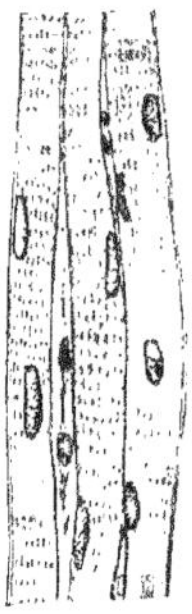

FIG. 15. Striated (skeletal) muscle of the rabbit.

*The position of the nuclei is characteristic of the so-called white muscles. In the semitendinosus of the rabbit, which is a red muscle, the nuclei occur between the fibril bundles of the interior of the fibre. Red fibres contract more slowly but are more resistant to fatigue. The proportion of red and white fibres varies in different muscles.

The muscular substance of the heart differs both from striated and smooth muscle in being composed of branched anastomosing fibres, which apparently form a continuous network. Like striated muscle, it possesses characteristic transverse markings, but, like involuntary muscle, it is under the control of the sympathetic nervous system.

As gross structures the voluntary muscles are functional units, each of which has a particular action according to the movement permitted by the parts of the skeleton to which it is attached. They present a longitudinal striation which is roughly referred to as the **direction of the fibres**, and which is of great value in identification. The striation is due to the circumstance that the fibres are arranged in parallel groups or muscle bundles, each of which is surrounded and separated from the adjacent bundles by a connective tissue enclosure, the **perimysium**.

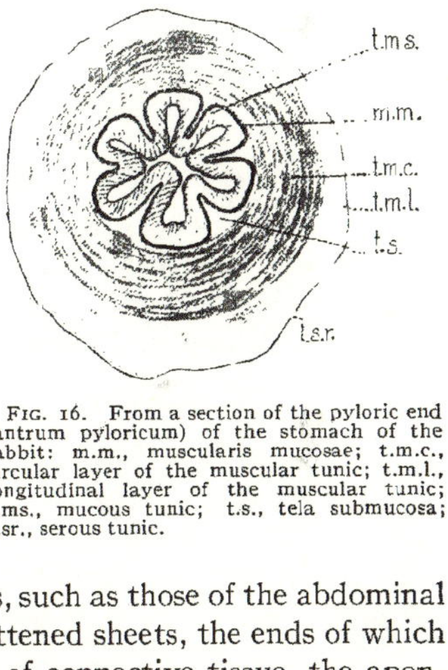

FIG. 16. From a section of the pyloric end (antrum pyloricum) of the stomach of the rabbit: m.m., muscularis mucosae; t.m.c., circular layer of the muscular tunic; t.m.l., longitudinal layer of the muscular tunic; t.ms., mucous tunic; t.s., tela submucosa; t.sr., serous tunic.

A **muscle** is typically spindle shaped, consisting of a middle fleshy portion, termed the **belly** of the muscle, and of tapering ends which provide for attachment. The attachment is effected by a strong band of fibrous connective tissue, the **muscle tendon** (Figs. 7, 34). Some muscles, such as those of the abdominal wall, are disposed in the form of flattened sheets, the ends of which are attached by broad, thin sheets of connective tissue, the aponeuroses. In **unipennate** muscles the fibres are attached obliquely to the side of the tendon, or in **bipennate** muscles to both sides, like the vane of a feather. In the so-called **biceps, triceps** and **quadriceps** muscles of the limbs, the origin is divided into two, three or four portions. The recognition of origin and insertion (p. 63) depends on usual but not invariable relations. The exact effect of muscle contraction depends as a rule on the relative positions of the parts and on the synchronous action of other muscles. A muscle like that forming the diaphragm does not

possess an insertion after the fashion of ordinary muscles; and in some cases, as in the intrinsic muscle of the tongue or the so-called **orbicular** or **sphincter** muscles, both origin and insertion may be absent.

Involuntary muscle is distinguished by its white or greyish coloration and by its smooth or homogeneous appearance. It forms characteristic layers in connection with visceral organs or with the skin, and is thus much less individual than the voluntary muscles in its relations to particular parts. It forms the muscular portion (**muscularis mucosae**) of the mucous tunic of the alimentary canal, and also a separate **muscular tunic** lying in the outer portion of its wall (Fig. 16.) In the muscular tunic the fibres are arranged in both circular and longitudinal directions. Involuntary muscle also forms a small constituent of many organs, such as certain glands, in which contractility is not a chief function. It forms a large constituent of the wall of the urinogenital tubes, particularly the bladder and the uterus. In association with elastic connective tissue it is an important constituent of the walls of the bloodvessels.

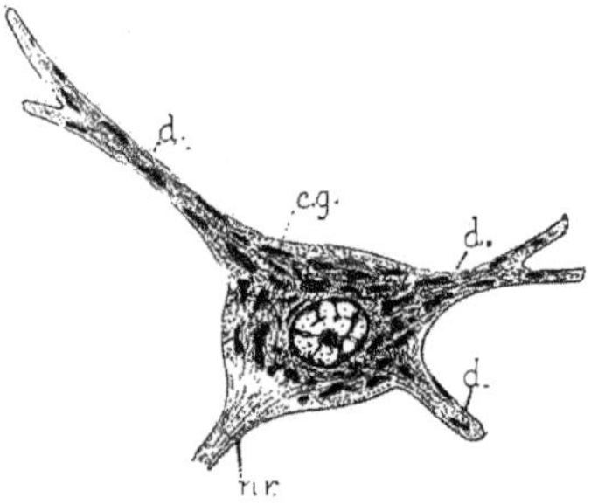

FIG. 17. Nerve-cell from the anterior grey column of the spinal cord (cf. Fig. 19): d., dendrites; c.g., chromatophile granules; nr., neurite.

Although there is an underlying community of structure in the walls of the bloodvessels, the two chief types of vessels, arteries and veins, present conspicuous differences, both in functional behaviour, and in their appearance in the dead animal. The differences are largely the result of differences in the relative amounts of the abovementioned constituents. The arteries are thick-walled, elastic tubes, which, under the force of blood from the heart, first become greatly expanded, and then gradually contract, so that the blood is forced into the smaller capillary vessels. The veins on the other hand are thinner-walled, less elastic vessels, through which the blood is forced largely through the pressure from behind. In the

dead animal the arteries appear white, flat or collapsed, and empty. The veins on the other hand appear large and dark on account of their distension with blood.

4. Nervous Tissues.

Nervous tissues form the basis of the central nervous system and of the outlying nerves and ganglia. They comprise two kinds of elements—**nerve cells** and **nerve fibres.** In the central nervous system these elements are imbedded in a mass of neutral tissue, the **neuroglia.**

Nerve cells are characteristic of the central nervous system and of the spinal and sympathetic ganglia. They differ greatly in form, but typically each consists of a **cell-body** (Fig. 17) bearing two kinds of processes—a fibre-process, the **neuraxis** or **neurite,** and a series of branched protoplasmic processes, the **dendrites.** The cell-body is distinguished by the presence in its interior of granular masses, the **chromatophile** or **tigroid** bodies. The latter extend into the dendrites, but not into the neuraxis. The dendrites may be greatly elaborated, and may be present to a considerable number. The neuraxis is a nerve fibre process. Since it continues as the central portion or axial cord of a nerve fibre, it may traverse a relatively enormous distance on its way to a peripheral organ.

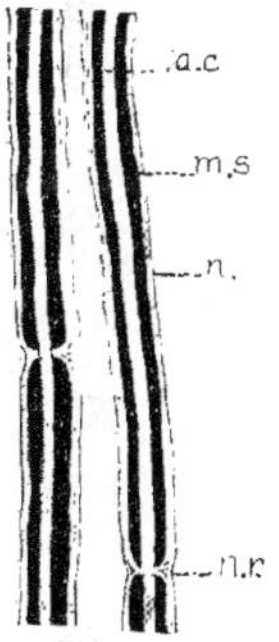

FIG. 18. Portions of two medullated nerve-fibres from the sciatic of the rabbit: a.c., axial cord; m.s., myelin sheath, stained black with osmic acid; n., neurilemma; n.r., node of Ranvier.

A nerve fibre consists of a central core, the **axial cord,** enclosed, except in the case of those of the olfactory nerve, by certain membranes. Two kinds of fibres are distinguished—**medullated** fibres, and **non-medullated** fibres. The former are characteristic of the peripheral nerves. In these (Fig. 18) the axial cord is surrounded by a comparatively thick membrane of fatty material, the **medullary** or **myelin sheath.** The latter is continuous except at certain points, the **nodes of Ranvier,** where the axial cord appears free except for an external investment of the whole fibre, the **neurilemma.**

In the non-medullated nerve fibres the myelin sheath is lacking. This type of fibre is chiefly distributed in connection with the sympathetic system.

A **nerve** is an association of nerve fibres, the latter being disposed in a parallel fashion and united together into bundles of larger or smaller size by connective tissue, which also forms a general peripheral investment, the **epineurium.** The dead-white coloration of a nerve is due to the fatty materials of the myelin sheaths, but nerves are commonly found imbedded in a fatty connective tissue which is associated with the epineurium and is also of white coloration.

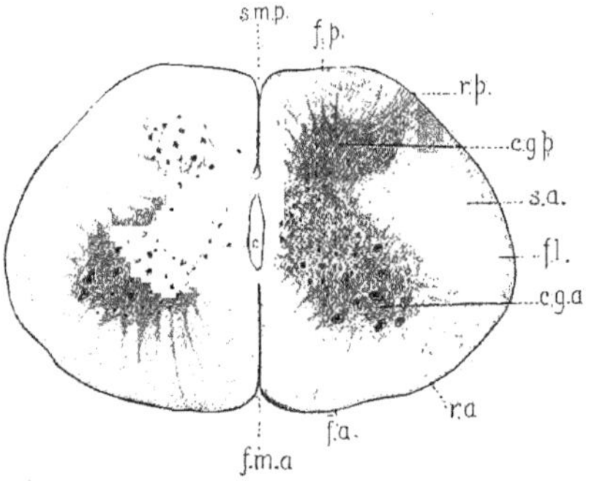

FIG. 19. Section of the spinal cord of the rabbit: c., central canal; f.m.a., anterior median fissure; s.m.p., posterior median sulcus; f.a., f.l., and f.p., anterior, lateral, and posterior funiculi of the cord; c.g.a. and c.g.p., anterior and posterior grey columns (horns of grey matter); r.a., and r.p., anterior and posterior nerve roots; s.a., white substance.

Nerve fibres, and also nerves, are distinguished functionally as **afferent** and **efferent.** They are organs of conduction, which carry impulses either from the peripheral parts of the body to the central nervous system, or in the opposite direction. **Sensory** nerves are afferent, while **motor** nerves aro efferent. Nerves, however, usually contain both afferent and efferent fibres and are hence described as **mixed.** In the distribution of afferent and efferent fibres there is a marked difference between the external or somatic portions of the body and the internal or visceral portions. Consequently, both **somatic** and **visceral** kinds of afferent and efferent fibres are conveniently distinguished.

On account of certain differences in coloration, the cellular and white fibrous constituents of the central nervous system produce characteristic patterns where either one is concentrated. Thus the cellular material is greyish, and is hence distinguished as the **grey substance,** while white fibrous material produces when concen-

trated an opaque white appearance similar to that seen in the peripheral nerves, and is hence described as **white substance.** In the spinal cord (Fig. 19) the grey substance is disposed as a central core, the white substance as a peripheral investment. A similar relation is found in the basal portion of the brain, but the characteristic pattern in the cerebral hemispheres and in the cerebellum is one in which the grey substance forms a peripheral, investing, or cortical layer.

5. Blood and Lymph.

Blood (Fig. 20) is fundamentally a cellular material, but owing to the fact that the cells or **corpuscles** are contained in a fluid medium, the **plasma**, it does not take on the features of the ordinary tissues of the body. It is for this reason also that, notwithstanding its important function, the appearances presented by the blood in dissection especially of preserved animals are negligible. The cellular materials consist of (a) **erythrocytes** or red blood cells, microscopic circular discs of fairly definite though not rigid contour, containing in the adult condition no nucleus. They have a yellow color, which is due to the presence of haemoglobin, and which gives to blood its deep red color when seen in bulk. The haemoglobin is the specific carrier of oxygen which it combines chemically. Arterial, oxygenated blood is bright red, while venous blood is dark red or bluish. The number of red blood cells is relatively somewhat greater in the rabbit than in man, there being over six millions contained in each cubic millimetre. The cells are formed in the vascular area of the embryo, later in the spleen and liver, and finally in the marrow of bones; (b) **leucocytes** and **lymphocytes**, also termed white or colorless blood cells. They are amoeboid, nucleated cells, present in much smaller numbers in blood, and in lymphatic vessels, and are formed in the lymph glands and spleen. They are capable of passing through the walls of the smaller vessels, and occur more or less throughout the tissues, where they have the function of carrying materials or of ridding the body of injurious substances and bac-

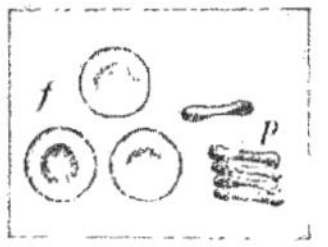

Fig. 20. Red blood corpuscles of the rabbit. f, face; p. profile.

teria; (c) **platelets,** minute nucleate amoeboid cells found in the blood stream.

In all multicellular animals the tissues are removed more or less extensively from the surfaces of absorption and excretion. Blood and lymph are the media through which tissue metabolism is maintained, and the vascular system is the mechanical means by which the continuous circulation of fluid is brought about. The composition of the blood varies from time to time according to the individual functions performed. Oxygen and food materials are carried to the tissues. Carbon dioxide and waste materials of other kinds are carried to organs from which they can be excreted. Blood, however, has been shown to vary in composition in different species of animals, and to be chemically homologous in related ones; while it possesses immunizing properties towards bacterial diseases, also differing in different species and individuals.

TERMINOLOGY.

In special or descriptive anatomy it is necessary to employ an extensive system of terminology in order that the various structures of the body may be individually designated, classified, and referred to their respective positions. The terms used for this purpose may be classified into four groups, as follows: (1) **general terms**—those included in the names of parts, but applicable in themselves to similar structures (arteries, nerves, etc.) in various parts of the body; (2) **specific terms** or **names of parts**; (3) **regional terms**—those defining areas (topographic); and (4) **terms of orientation.**

Although few in number, the terms of orientation may be regarded as the most generally useful terms of descriptive anatomy. This is because they are based on very general relations of the body and are therefore of wide application. For this reason they are here selected for definition to the exclusion of others of a more restricting or individualizing kind.

In all vertebrates we may recognize a **longitudinal axis,** corresponding, in general, to the line formed by the vertebral column. In the usual or **prone** position of the body this axis is horizontal. The uppermost surface is then described as **dorsal,**

the lowermost surface as **ventral**, the sides of the body as **lateral.** Any position forward, with respect to the long axis, is **anterior** in comparison with any position backward, which is **posterior.**

In relation to the long axis it is convenient to recognize a **median vertical plane** of section, which is one dividing the body into right and left halves; also **transversal planes**, which are planes situated at right angles to the median plane and to the long axis, and **sagittal planes.** The latter are vertical planes parallel to, and also including (as midsagittal), the median vertical plane.

The median vertical plane is the plane of **bilateral symmetry,** each half of the body as thus defined being in a general way the reverse counterpart of the other. Structures situated in part in the median plane are unpaired, and are described as **median**, while structures situated wholly outside of the plane are paired, right and left, or **dextral** and **sinistral.** In relation to the median plane and to the sides of the body, structures are described as **medial** when nearer the former, and as **lateral** when nearer the sides of the body. The term **intermediate** is applied especially to a position between medial and lateral, but this restriction is perhaps not justifiable.

In considering the extent of bilateral symmetry, it is necessary to bear in mind that, although a fundamental feature in vertebrates, it is not perfectly retained in the adult condition. Symmetry is destroyed by the migration of an unpaired structure from a median to a lateral position, as is seen, for example, in the case of the abdominal portion of the alimentary canal; or, again, by the reduction or disappearance of structures belonging to one side of the body, as, for example, in the case of the mammalian aortic arch.

Referring to centre and circumference, either in the body as a whole, or in particular parts, the terms **deep** and **superficial, central** and **peripheral,** or **internal** and **external** may be applied. It may be observed, however, that the terms internal and external are sometimes used in the sense of medial and lateral, both in descriptive language and in the names of parts.

In comparison with the terms medial (medialis) and median (medianus) the term **middle** (medius) may be used to designate the position of a structure lying between two others, the latter being otherwise designated, for example, as anterior and posterior, or one in the median plane.

The limbs being more or less independent structures, it is proper to apply to them certain terms otherwise applicable to the main portion of the body. The chief terms not properly applied elsewhere are **proximal**, meaning nearer the centre or base of attachment, and **distal**, toward the extremity. In the middle segment of the fore limb the respective positions of the bones (radius and ulna) are indicated as **radial** and **ulnar.** The terms **tibial** and **fibular** are also applicable, although with less reason, to the corresponding segment of the hind limb. The dorsal and ventral surfaces of the fore foot are described respectively as **dorsal** and **volar,** those of the hind foot as **dorsal** and **plantar.**

In determining the identity of structures in a quadrupedal mammal, considerable difficulty may at first arise from the fact that descriptive terms, such as those just defined, are frequently included in the names of parts, the latter being, at the same time, terms applied in the first instance to the human body, in which the recognized relations are somewhat different. In comparison with that of a quadrupedal vertebrate, the human body occupies a vertical or **erect** position, and is to be considered as having been rotated upward through ninety degrees on the posterior limbs. The latter accordingly occupy for the most part their original position, and the human arm largely reassumes this position when allowed to hang freely at the side of the body. As in all cases, the face retains its forward direction. Thus the terms anterior and posterior as used in human anatomy mean ventral and dorsal, provided they refer to parts of the body, such as the entire trunk region, which have been affected by rotation. The terms **superior** and **inferior** as applied to man are similarly anterior and posterior as applied to a lower form. Since it is unwise to change the form of the official terms of human anatomy, it becomes necessary to interpret all such terms when used for a quadrupedal mammal according to the relations exhibited by man. The human terms may in most cases be translated into terms acceptable for comparative anatomy by reading ventral for anterior, dorsal for posterior, **cranial** or **oral** for superior, and **caudal** or **aboral** for inferior. The exceptions then apply to those parts of the body unaffected by rotation.

On the other hand in ordinary description of organs and their position, where it is not a matter of the official names of parts, little advantage is to be gained from adherence to this principle. The terms anterior and posterior apply with much greater force to a lower vertebrate, than to man, while the terms superior and inferior are only of interest in the latter. In this case the rule here followed is to use the terms anterior and posterior *for descriptive purposes* without reference to the human relation. The same remark applies to the terms of direction, *viz.*, upward, downward, forward, and backward.

In this connection it may be pointed out that the custom has become more or less general in comparative anatomy of employing the termination **ad** with words otherwise signifying position alone, in order to indicate position or course toward, *e.g.*, dorsad = dorsalward. In the present case this form is used only for course, position being indicated by the adverbial termination **ly**, *e.g.*, dorsally.

Reference may also be made here to the fact that the human structures to which identifying names are applied frequently fail in one way or another to correspond to structures in a lower form. Composite structures to which individualizing names are applied, for example, may be represented by independent parts. Also, structures which are similar in form or function may be convergent. Finally, although it is essential to endeavour to apply all terms as accurately as possible, it will be remembered that a terminology primarily arranged for one type cannot be exactly applied to another without considerable qualification.

THE GENERAL FEATURES AND GROUND PLAN OF THE ORGAN SYSTEMS.

It has already been stated as a general principle that the structure of an organism is the expression of an underlying plan and pattern, in the elaboration of which embryonic development and ancestry play a very large part. How, as a matter of interpretation, the comparative method is applied, may be demonstrated by reference to any part of the body of an animal; and in the following pages will be found, under the head of the respective systems, a

preliminary statement of how certain outstanding features of the rabbit may be explained, as to the nature of their origin, and what grades of organization they may be presumed to illustrate. It will be recognized that the mammals are in most respects less specialized than man, and must accordingly show in their grades of organization various stages through which man must be assumed to have passed.

CLASSIFICATION OF THE ORGAN SYSTEMS.

The term **organ-system** is employed in descriptive anatomy to designate a group of organs which cooperate in a general function. In many respects the systems represent primitive functions, and it is therefore largely on account of the independent elaboration of these that the systems may be recognized also on a structural basis as groups of organs allied in origin and development. The exact number of systems recognized depends on certain arbitrary distinctions, the following being those usually distinguished:

(1) The **integumentary system,** comprising the **skin,** and its appendages, namely, the **hairs** and the **general cutaneous, mammary,** and **inguinal glands.**

(2) The **skeletal system,** comprising the cartilage and bone elements of the skeleton, with their connections.

(3) The **muscular system,** comprising all contractile structures of the body. Since, however, the involuntary muscles are arranged as muscle layers in connection with visceral organs, the muscular system is usually considered as including only the individual or voluntary muscles of the skeleton and skin.

(4) The **nervous system,** comprising the **central nervous system** (the **brain** and **spinal cord**) and the **peripheral nervous system,** the latter consisting of the paired **cranial** and **spinal nerves** with their associated ganglia. A special portion of the peripheral nervous system is set apart as the **sympathetic nervous system.** The latter consists of a pair of ganglionated **sympathetic trunks** lying along the ventral surface of the vertebral column, and of two series of ganglia, **prevertebral** and **peripheral,** connecting the trunks with the visceral organs.

(5) The **digestive system,** comprising the **digestive tube** and its outstanding glandular appendages—the **oral glands,** the **liver,** and the **pancreas.**

(6) The **respiratory system**, comprising the **lungs**, and **respiratory passages**, namely, the **bronchi**, the **trachea** and the **larynx**. With this system may also be included the **accessory respiratory passages** formed by the **nasal fossae.**

(7) The **vascular system**, comprising the organs of circulation,

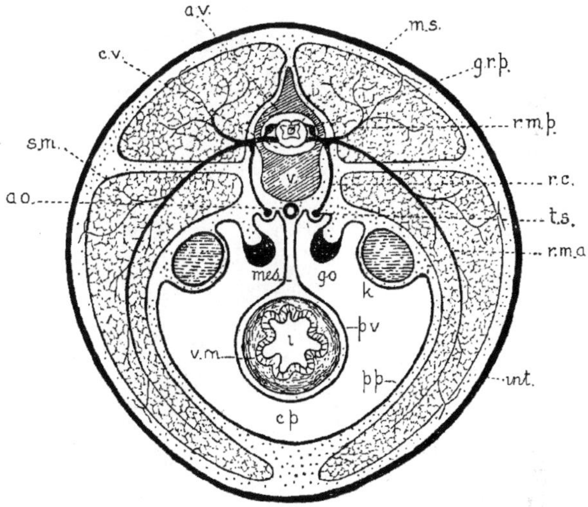

FIG. 21. Schematic representation of the chief organ-systems of a generalized vertebrate as seen in a transverse section of the abdominal region:

Integument—int.

Skeleton—v., vertebral body; a.v., vertebral arch; c.v., vertebral canal.

Muscular system—s.m., skeletal muscle; v.m., visceral muscle.

Nervous system—m.s., spinal cord, with the central canal, and the dorsal (posterior) and ventral (anterior) roots of the spinal nerves; g.r.p., ganglion of the posterior root; r.c., ramus communicans to sympathetic trunk; r.m.a. and r.m.p., anterior and posterior rami of a spinal nerve; t.s., sympathetic trunk.

Digestive system—i. intestine.

Vascular system—ao., aorta.

Urinogenital system—k., kidney; go., gonad (ovary or testis).

Serous cavity—c.p., general coelom, pleuroperitoneal, or peritoneal cavity; p.v. and p.p., visceral and parietal parts of the serous tunic—visceral and parietal peritoneum; mes., mesentery.

namely, the **heart, arteries, capillary vessels,** and **veins.** The lymph-conducting canals are also portions of the circulatory system, but since they are largely independent of the bloodvessels, they are usually considered as forming with their associated lymph glands a separate **lymphatic system.**

(8) The **urinogenital system**, comprising the reproductive and excretory organs, together with their common ducts—the **urethra** of the male and the **vestibulum** of the female—and the associated **bulbourethral gland.** The reproductive organs comprise, in the male, the central organs or **testes**, and the **deferent ducts**, both of which are paired, the unpaired **seminal vesicle,** and the paired **prostatic** and **paraprostatic glands.** In the female, the reproductive organs comprise the paired **ovaries, uterine tubes** and **uteri,** together with the unpaired **vagina.** The excretory organs of both sexes comprise the paired **kidneys** and **ureters** and the unpaired **urinary bladder.**

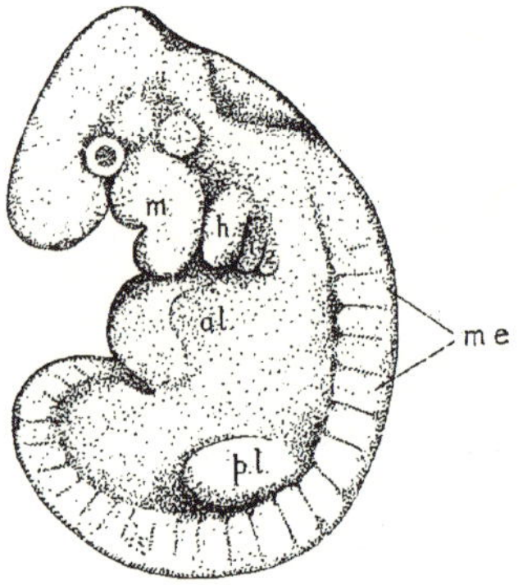

FIG. 22. Rabbit-embryo of 10½ days (4.8 mm.): m., mandibular; h., hyoid; 1 and 2, first and second branchial arches; a.l., anterior limb-bud; me., metameres; p.l., posterior limb-bud. (After Minot and Taylor, in Keibels Normentafeln, V.; Fig. 12.)

Certain organs of the body are not included in this classification: (1) The **thymus** and **thyreoid glands** are connected with the digestive tube in the embryonic condition, and for this reason are sometimes included with the digestive system, although in the adult they occur as independent structures. (2) The **suprarenal body** is situated close to the kidney of either side, but is independent of the latter, both in the adult condition and in point of development. (3) The **special** (olfactory, optic and auditory) **sense-organs** of the head are highly elaborated structures, the relations of which are partly with the central nervous system.

GENERAL ORGANIZATION.

In the rabbit, as in all vertebrates, the general plan of organization involves three chief features, as follows:

(1) **Axial orientation.** The axial line of the body is indicated by the chief portion of the skeleton, and the important organ-

systems are grouped in a characteristic fashion about it (Figs. 21, 23).

(2) **Metamerism.** A large portion of the body, mainly dorsolateral in position, is arranged on a segmented plan, in which parts are repeated serially and longitudinally around and to either side of the original axis. Metamerism does not appear to any extent on the surface of the adult body, but becomes evident in

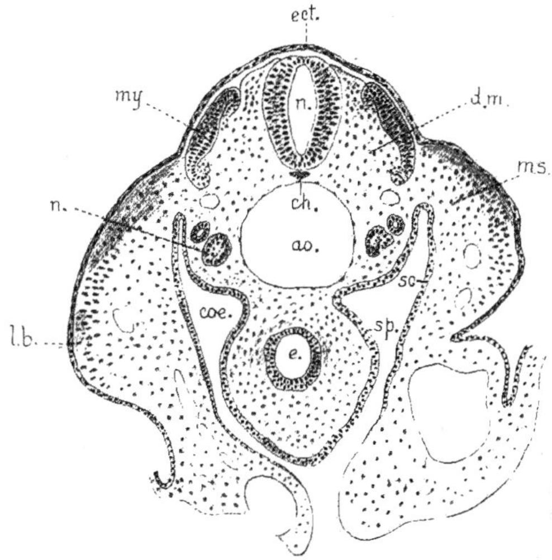

FIG. 23. Transverse section of a rabbit-embryo of about 10½ days, showing the arrangement of the organ-systems: ao., aorta; ch., notochord; coe., coelomic cavity; d.m., dorsal mesoderm (myotomic and sclerotomic divisions); e., primitive alimentary canal (enteron); ect., ectoderm; l.b., limb-bud; ms., mesenchyme; my., external portion of a myotome; n, nephrotome of embryonic kidney; intermediate mass of mesoderm; sp. and so., splanchnic (visceral) and somatic (parietal) portions of the ventral mesoderm.

structure in the subdivision of the vertebral column into vertebrae and the paired, serial arrangement of the related spinal nerves, vessels and musculature. Metamerism is externally evident in embryos (Fig. 22) and is anatomically founded upon the serial arrangement of parts of the mesoderm.

(3) **Branchiomerism.** This is an adult feature of lower aquatic vertebrates (Fig. 30), such as fishes, where it appears as a

succession of true gill or branchial structures, which support gill filaments as functional respiratory organs. In higher terrestrial animals it appears as an embryonic feature (Fig. 22, m.h. 1,2) and is to be considered both as a determinant of adult form and as a mark of aquatic ancestry. As in lower vertebrates, it underlies not only structures of branchial significance (branchial arches in the restricted sense, but also modified branchial structures, including the first or mandibular arch (m.), and the second or hyoid arch (h.). So great is the modification of these structures in passing from the embryonic to the adult condition that the recognition of the ground-plan is perhaps here of less general importance. It, however, determines the position and relations of certain skeletal structures, including the auditory ossicles, the hyoid, and in part the laryngeal cartilages—a point of some value in the classification of the parts of the head skeleton. It also determines the succession of certain soft structures, including the fifth, seventh, ninth and tenth cranial nerves; also the chief arterial vessels of the heart, which are more fully referred to below.

The fundamental importance of branchiomerism lies in the fact that respiration by means of gill perforations of the pharynx is characteristic of that branch of the animal kingdom from which all chordate animals have been derived. In the various invertebrate phyla are found respiratory surfaces of many kinds, in which, however, thin surface membranes, external tufted, or invaginated tubes and analogous structures, provide for diffusion without perforation of the body-tube.

EMBRYONIC PLAN OF THE SYSTEMS.

In the individual organ-systems the main features of the general plan, as estimated on embryonic development, may be outlined as follows:

1. The formation of an axial skeletal support, consisting primarily of a strand of cellular tissue, the notochord, and secondarily of a segmented cartilaginous, afterwards bony, vertebral column.

2. The formation of (a) a primary cartilage skull (chondrocranium) as a support for the brain and capsules of the special sense organs (neurocranium or cerebral cranium); and (b) a series of cartilaginous visceral arches (splanchnocranium or visceral cranium).

3. The formation of the chief skeletal muscle in a dorsolateral position along the axis.

4. The formation of the central nervous system as a tube of nervous matter (neural tube), lying on the dorsal side of the axial support, and differentiated into a general posterior portion, the spinal cord, and an anterior expended portion, the brain.

5. The formation of the digestive tube as a median structure, lying directly beneath the axial support, and of special glandular appendages arising from the epithelium of its wall.

6. The formation of the lungs as paired outgrowths of the ventral wall of the digestive tube, afterwards connected with the outside of the body by accessory respiratory tracts traversing the head.

7. The formation of the circulatory system primarily on an aquatic plan. This involves the formation of (a) the heart in a ventral position to the digestive tube and immediately behind the gills; (b) a ventral aorta, passing forward to the gills, and dividing into a paired series of branchial aortic arches; (c) a dorsal aorta, combining the aortic arches, and passing backward along the ventral surface of the axial support; and (d) a series of paired veins returning the blood from various parts of the body to the heart.

9. The formation of the reproductive organs or gonads in association with the dorsal lining of the coelomic cavity, and their connection with the outside of the body by modified kidney ducts.

10. The formation of the kidneys, either as embryonic or permanent structures, from an intermediate mass of tissue, lying in general between the dorsal musculature and the lining of the coelomic cavity (cf. position of embryonic kidney in Fig. 23).

11. The formation, in the ventral portion of the body, of an extensive space, the coelomic cavity or coelom, afterwards differentiated into pericardial, pleural and peritoneal portions.

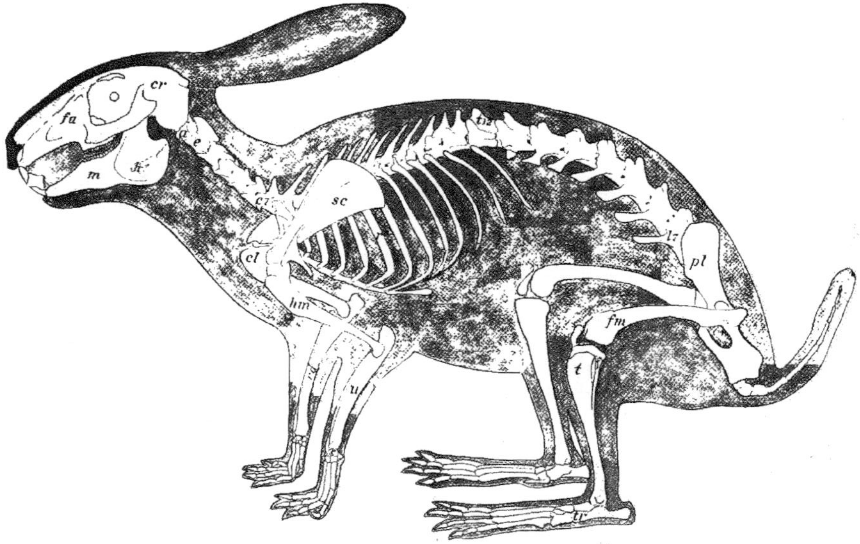

FIG. 24. The skeleton: a, atlas; c, carpus; c7, seventh cervical vertebra; cl, clavicle; cr, cranial portion of skull; e, epistropheus; f, fibula; fa, facial portion of skull; fm, femur; h, hyoid; hm, humerus; l7, seventh lumbar vertebra; m, mandible; pl, pelvis (coxal bone); r, fifth rib; rd, radius; t, tibia; t12, twelfth thoracic vertebra; tr, tarsus. The position of the clavicle shown above its designation.

THE SKELETAL SYSTEM.

The designation "vertebrate" has reference to a common feature of higher animals, fishes, amphibians, reptiles, birds and mammals—the possession of a backbone or **vertebral column,** composed of individual segments, the **vertebrae.** Vertebrates are, however, more properly described as animals having an internal skeleton.

VERTEBRATE ANIMALS.

The vertebral column is part of the principal, **axial skeleton** (Fig. 24), otherwise formed by the ribs, sternum, and skeleton of the head. In addition to the axial skeleton, nearly all vertebrates possess an **appendicular** part of the skeleton devoted to the support of the limbs, or in terrestrial vertebrates, more correctly speaking, to the support of the body on the limbs, and to locomotion. In both fore and hind limbs, this consists of a proximal portion lying within the contour of the body and forming the **pectoral** and **pelvic girdles**, and of a distal portion, lying beyond the general contour of the body and comprising the **skeleton** of the **free extremities.** The limbs of vertebrates present an extraordinary range of adaptations, being modified in the various groups into fins, paddles, wings, and walking or running limbs. In the majority of cases their adherence to a common ground plan is evident from their composition.

The vertebral column of the rabbit consists of 7 **cervical,** 12 **thoracic,** 7 **lumbar,** 4 **sacral,** and 14-16 **caudal** vertebrae. The vertebrae are found to be gradually modified from any intermediate part of the column forward or backward, but characteristic vertebrae can be identified for each region. The vertebrae are not arranged in a straight line; on the other hand, there is a curvature dorsad in the thoraco-lumbar region, and corresponding curvatures ventrad in the cervical and caudal regions. In fishes the line of the vertebral column is straight, and there is little indication of regional differentiation. The feature of curvature appears in terrestrial vertebrates, especially mammals, and is related to the support of the body in a lighter medium. The curvatures are modified in man, in view of the erect position, in that there is a lumbar curvature ventrad counteracting the effect of the dorsal curvature of the thorax, while the caudal

VERTEBRAL REGIONS.

region is reduced to a vestigial **coccyx**, consisting of coalesced vertebrae.

The individual vertebra is made up of a more massive ventral **body**, upon which is set a bridge of bone forming the **arch**. Both

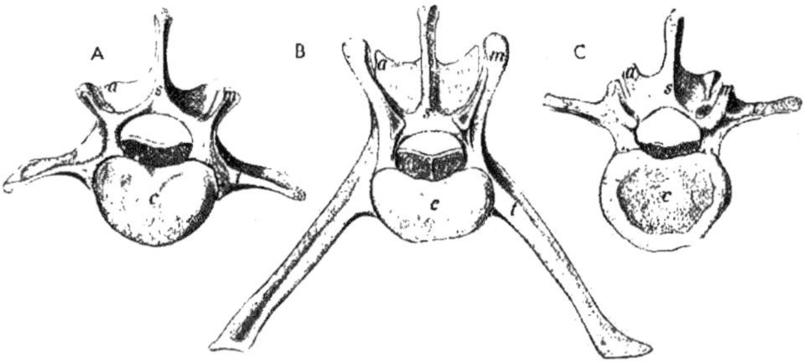

FIG. 25. Mid-lumbar vertebrae of bear (A, fifth), rabbit (B, fifth), and man (C, third): a, inferior articular process; c, body; m, mamillary and superior articular process; s, spinous process; t, transverse process.

body and arch bear **processes** which reach into the surrounding muscles, and serve for their support. The principal processes are

THE FORMS OF VERTEBRAE.

mid-dorsal or **spinous**, lateral or **transverse**, and, in the lumbar region, dorsolateral or **mamillary**. Generally speaking every vertebra has three principal functions—to support the body, to protect the spinal cord, and to offer support for muscles.

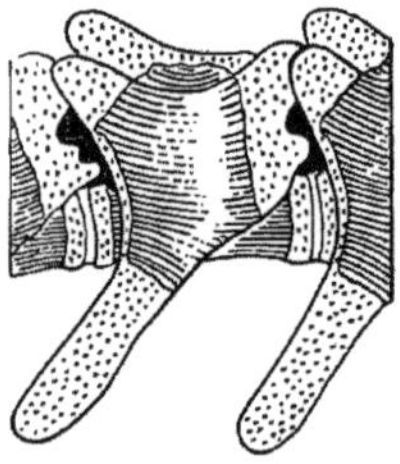

FIG. 26. Lumbar vertebra of four-day-old rabbit. Cartilage dotted, bone shaded. X 5.

Any vertebra of the rabbit may be compared with the corresponding vertebra of any mammal or in a general way with those of any vertebrate. The fifth lumbar vertebra of the rabbit, for example, would be found distinguished by the great development of its processes, since the latter support powerful muscles used in leaping. The corresponding human vertebra, or the third as functionally more nearly equivalent, is weak in muscular expansion, but its body is massive for purposes of

axial support. A corresponding vertebra of the bear will be found more or less intermediate between the two types (Fig. 25).

The adult vertebra of the rabbit and of higher vertebrates in general is composed of bone. It arises, however, embryonically in cartilage (Fig. 26). The transition from cartilage to bone is based on the primary condition in lower vertebrates in which the entire skeleton, vertebrae included, is formed in cartilage, and may remain in this condition throughout life.

CARTILAGE AND BONE.

The **axial line** of the vertebrae in mammals passes through the central portions of the bodies. This position is marked, in the embryonic condition only, by the **notochord** (Fig. 23). Some of the lower aquatic vertebrates, such as lampreys, exhibit the notochord in both young and adult conditions, and show little indication of the development of the elements of vertebrae. Others of slightly more advanced position, such as sharks (Fig. 27), show the notochord, extending more or less to the adult condition with the vertebral elements developed round about it.

THE NOTOCHORD.

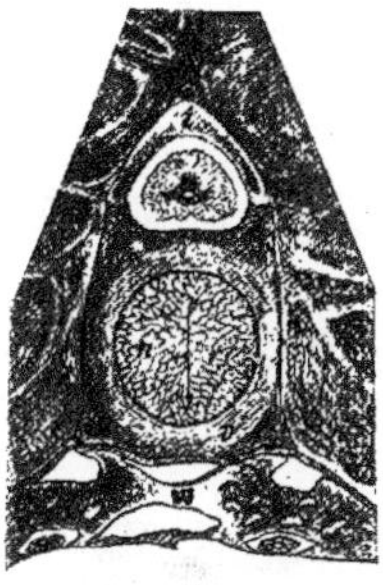

FIG. 27. Transverse section of shark vertebra (cartilage stage), from young specimen of Atlantic dogfish, *Acanthias*. i, intercalary cartilage, completing arch; n, notochord; v, body of vertebra.

The head skeleton of a mammal, usually but inaptly called the skull, is a complex of individual bones and cartilage, the arrangement and functions of which may be determined with a little effort. The general disposition of the bone elements, demonstrable in the rabbit or any mammal is as indicated in Fig. 28. Briefly, there is a linear series of basal segments, comprising from behind forward **basioccipital, basisphenoid, presphenoid, and mesethmoid.** The three first-named form the floor of the brain-case, while the mesethmoid forms the nasal septum. Associated with the basioccipital are paired, lateral **exoccipital** bones, and a **supraoccipital** element, together forming an occipital ring en-

COMPOSITION OF THE SKULL.

closing the aperture for transmission of the spinal cord from the cranial cavity. The basisphenoid and presphenoid bear lateral expansions, respectively the **greater** and **lesser wings**, or **ali-sphenoids** and **orbitosphenoids,** which assist in the formation of the side walls of the brain case. The bone capsule (**periotic**) lodging the internal ear on either side is solidly built into the lateral cranial wall between the exoccipital and alisphenoid, while further forward the light scroll-like surfaces of the **ethmoid bone,** or **ethmoturbinal,** representing the chief portions of the **nasal**

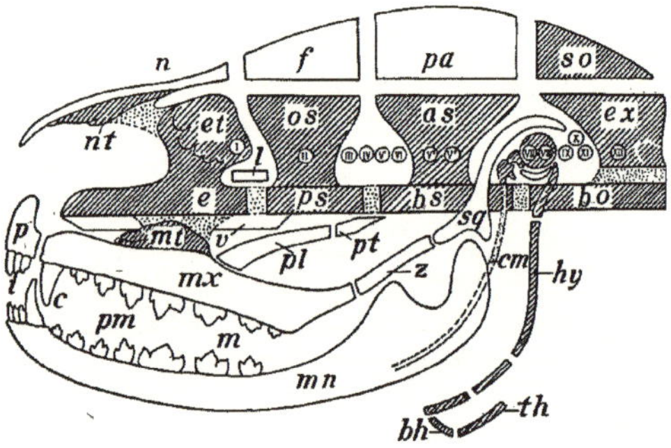

FIG. 28. Composition of the mammalian skull. Cartilage dotted, cartilage bone shaded, derm bones plain. I-XII, cranial nerves; as, alisphenoid; bh, body of hyoid; bo, basioccipital; bs, basisphenoid; c, canine teeth; cm, Meckel's (mandibular) cartilage (I); e, mesethmoid; et, ethmoturbinal; ex, exoccipital; f, frontal; hy, hyoid (II); i, incisors; l, lacrimal; m, molar teeth; mn, mandible; mx, maxilla; mt, maxilloturbinal; n, nasal; nt, nasoturbinal; os, orbitosphenoid; p, premaxilla; pa, parietal; pl, palatine; pm, premolar teeth; pt, pterygoid; sq, squamosal; so, supraoccipital; th, thyreohyal (III); v, vomer; z, zygomatic.

capsule is attached on each side of the base of the mesethmoid The original proportions of the nasal capsule are however more nearly commensurate with the general cavity of the nose; and additional turbinal surfaces, comprising the **nasoturbinals** and **maxilloturbinals** are attached secondarily to the nasal and maxillary bones.

To this foundation there is added a series of enclosing bones, for the most part thin and superficial, but nevertheless making up the greater part of the **facial** portion of the head skeleton as opposed to the **cranial** or brain containing portion. The series comprises the **interparietal** (absent in the hare and many mam-

mals), paired **parietals, squamosals, frontals,** and **nasals** as roofing bones, and paired **premaxillary, maxillary, palatine, pterygoid,** and **mandibular** bones forming the solid supports of the mouth. Several lateral elements also take part, including the **lacrimal,** at the anterior border of the orbit, the **zygomatic,** forming the central portion of the corresponding arch, for protection of the orbit and muscular support, and finally the bladder-like **tympanic** bone, which forms the enclosure of the middle ear and protects the delicate bones of the **auditory chain.**

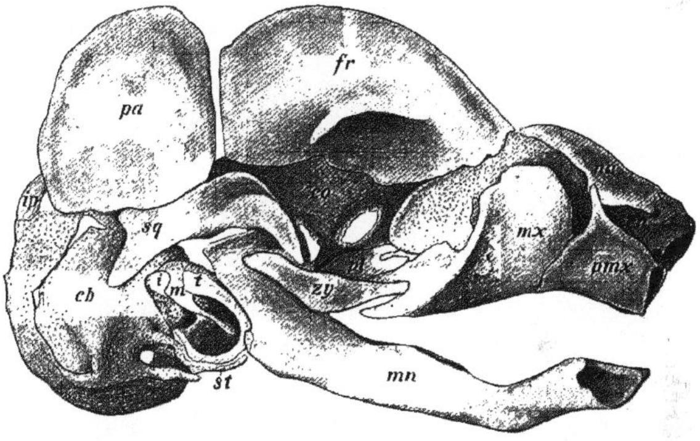

FIG. 29. Lateral view of skull of rabbit foetus, 45 mm. cb, co, cn, cranial, orbital, and nasal portions of primary chondrocranium; fr, frontal; i, incus; ip, interparietal; m, malleus; mn, mandible; mx, maxilla; na, nasal; pa, parietal; pl, palatine; pmx, premaxilla; sq, squamosal; st, styloid process; t, tympanic; zy, zygomatic; Born plate model, after Voit.

An important though inconspicuous portion of the head skeleton is formed by the **hyoid apparatus** supporting the tongue, and certain cartilages of the larynx, with which the hyoid is intimately associated. The relation of this complex to the skull is indicated in a mammal by the suspension of the hyoid apparatus from its base. The **malleus, incus** and **stapes** of the auditory chain form with the elements just described that portion of the visceral skeleton as modified in the mammalian skull, with the exception, as described below, of certain replacing or derm elements also considered to belong to this division.

The skull consists primarily in the embryo of a cartilage trough, the extent of which is roughly definable as the area occupied by the occipital, anterior and posterior sphenoidal, and ethmoidal portions (Fig. 29.) As a cartilage skull it is designated as the **chondrocranium**, and after its conversion into bone as the **osteocranium.** It is no more than an enclosure for the brain, except that it has associated with it the cartilage capsules of the nasal, visual, and auditory organs, and, in the case of the first and last of these, the capsules are incorporated with the skull proper. Thus, the primary skull is designated as the **neurocranium** or **cerebral cranium,** to distinguish it from a second portion of the head skeleton, the **splanchnocranium** or **visceral cranium,** which includes the series of visceral arches suspended from the ventral surface of the neurocranium. The addition to the primary head skeleton of a large number of membrane bones results in more or less confusion of the original divisions, since the membrane portions of the visceral cranium are, with the exception of the mandible, united by suture with those of the cerebral cranium, while the true cartilage or cartilage bone portions of the former, occurring as the auditory ossicles, the hyoid and larnyx (in part), although highly modified, remain in a more or less independent relation.

CHONDROCRANIUM AND OSTEOCRANIUM.

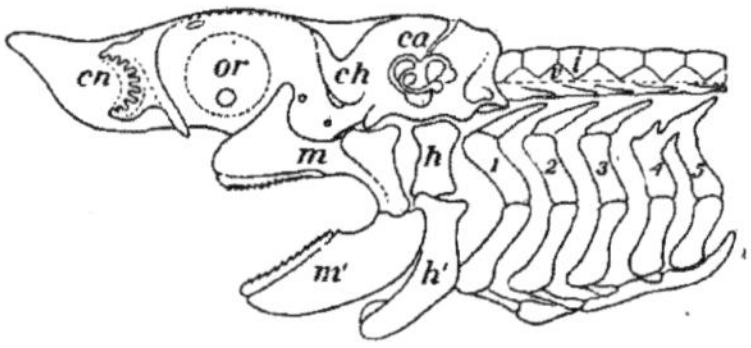

FIG. 30. The chondrocranium and visceral arches of the Atlantic dogfish, *Acanthias*: ca; auditory capsule; ch, chondrocranium; cn, nasal capsule; h,h', dorsal and ventral segments of hyoid arch (II); i, intercalary cartilage of vertebral column; m,m', dorsal and ventıal poıtions of mandibular arch (II), functional upper and lower jaws; malleus and incus of mammalian ear; or, orbit, depression for optic capsule; v, vertebra; 1-5, branchial arches.

The appearance of the mammalian skull during the later stages of foetal development is most striking, the cartilage mass of the chondrocranium, and the bones ossifying in its interior forming a foundation basal mass, from which are suspended elements of the same nature, principally auditory and hyoid, in a somewhat arch or rod like form. The auditory arch is formed by the two more lateral bones of the auditory chain, incus and malleus, of which the

incus is attached to the skull, while the malleus is extended as the **cartilage of Meckel** almost the whole length of the inner surface of the mandible. The bulk of the skull is formed already by the surface elements distributed in the characteristic fashion, but as yet only loosely associated (cf. Fig. 29).

This condition of the developing skull in a mammal finds its explanation far back in the history of the vertebrates and is only made clear by the study of some one of the lower fishes such as shark or sturgeon.

HEAD SKELETON OF LOWER VERTEBRATES.

In a shark (Fig. 30) the entire internal skeleton is formed in cartilage which persists throughout life. The principal part of the head skeleton is a massive cartilaginous box (chondrocranium), enclosing the brain, and including as a result of growth and fusion the nasal and auditory capsules. The eye capsules are free, and their accommodation at the side of the cranium establishes the orbital depression to be seen on the skull of all vertebrates. This type of structure is obviously the basis of the mammalian skull, except that in the latter the cartilage mass is at once more restricted to its basal portion, and is more specialized by its replacement by definite bone centres.

FIG. 31. Shagreen denticles of the smooth dogfish, *Mustelus;* enlarged.

It is, however, in respect of the visceral arches suspended from the chondrocranium that the structure of the shark skull is most illuminating. The first or **mandibular** arch forms the upper and lower jaw. Its composition reflects that of the external part of the auditory chain in the mammalian embryo, and establishes a fact, for a long time scarcely believed by zoologists, that in the history of mammals this arch has undergone a profound change of function. The second or **hyoid arch,** though developed to an extent out of keeping with the degree of development of the primitive tongue, is nevertheless obviously homologous with the hyoid arch of a mammal and in most sharks has an important accessory function in the support of the lower jaw. Following the hyoid arch are five ordinary or **branchial arches** supporting the filaments of the gills and serving as pillars of the gill apertures. These arches are then the parts of an extensive system out of which those portions of the laryngeal cartilage which are of visceral rather than tracheal origin have been specialized in mammals.

So far as the surface or roofing portion of the skull is concerned, it is represented in a shark only by dermal teeth or shagreen denticles (Fig. 31), uniformly distributed in the skin of the body, and concentrated at the aperture of the mouth into definite teeth lining the jaws. In sturgeons and related fishes, however, these structures are already concentrated into a definite pattern of surface plates, having in general the same disposition as derm elements in the skull of all higher vertebrates, and distributed in such a way that they form an almost complete enclosure for those parts of the chondrocranium which they invest. The pattern of the dorsal elements is best indicated in fossil amphibia and early reptilia, in which the plan is almost diagrammatic.

DERM BONES.

The elements of the head skeleton may be classified as follows:

1. The **CEREBRAL CRANIUM** (cranium cerebrale or neurocranium), including:
   - (a) The primary cartilage skull (chondrocranium), enclosing the brain, and containing in its wall the olfactory and auditory capsules (embryonic);
   - (b) The secondary bone skull (osteocranium), replacing (a) and comprising the **occipital, sphenoid, ethmoid, inferior turbinal,** and **periotic** bones;
   - (c) The associated derm elements, comprising the **interparietal, parietal, frontal, nasal, vomerine, lacrimal, tympanic,*** and **squamosal** bones.

2. The **VISCERAL CRANIUM** (cranium viscerale or splanchnocranium), including:
   - (a) The primary mandibular and hyoid† visceral arches (embryonic);
   - (b) The secondary elements, replacing (a)—the **malleus, incus,** and **stapes** of the auditory chain; the **hyoid** bone and its connections with the skull;
   - (c) The associated derm elements of the face and palate, comprising the **premaxillary, maxillary, zygomatic, mandibular, palatine** and vestigial **pterygoid** bones.

---

*The identification of the tympanic as a derm element has been questioned.

†The thyreoid cartilage of the larynx and its connection with the hyoid (greater cornu) are modified branchial arches, but the structure as a whole is not included with the head skeleton.

As accessory structures the **teeth** of the rabbit present two mammalian features; they are **heterodont,** or differentiated according to particular regions; and the adult series, excepting those designated as molars, are **permanent** teeth, replacing **deciduous,** or **milk** teeth of the young animal. The condition is thus described as **diphyodont** in comparison with that in lower vertebrates, where there is usually a multiple tooth change, new teeth being developed as required (polyphyodont type).

DENTITION.

Moreover, in the rabbit, as in all mammalia, the number is restricted, so that, considering the differentiation of the teeth, it is possible to express their relations by a **dental formula.** In the mammalia generally the teeth are differentiated into **incisors, canines, premolars** and **molars,** and in placental mammals the full dental formula is indicated as i. $\frac{3}{3}$, c. $\frac{1}{1}$, pm. $\frac{4}{4}$, m. $\frac{3}{3}$. In the rabbit as in other rodents, however, the dentition is greatly modified by the elaboration of two pairs of incisors for gnawing and the corresponding obliteration of intermediate teeth, the place of the latter being occupied by an extensive gap, or diastema, in which no teeth occur. The dental formula of the rabbit is i. $\frac{2}{1}$, c. $\frac{0}{0}$, pm. $\frac{3}{2}$, m. $\frac{3}{3}$. It will also be observed in this animal that the absence of the intermediate teeth allows the lips to be approximated behind the incisors, and since in this region the lips are also provided with hairs on their internal surfaces, the oral cavity is separated almost completely from a small space enclosing the incisor teeth. This adaptation, however, is not so perfectly developed in the rabbit as in certain others of the rodent order.

The human skull differs most markedly from that of the rabbit or other mammal in the enormous development of the cranial region, and in the anteroposterior compression of the face, with which is associated, shortening of the jaw region, reduction of the nasal cavities and rotation of the orbits to a forward position.

THE HUMAN SKULL.

A most instructive feature is the re-adjustment of the axis, coincident with the assumption of the erect position. In both quadrupedal and bipedal positions the face naturally retains its forward direction. In most mammals, as in vertebrates generally, the axial line of the cranium, known as the **basicranial axis,** and that of the face, the **basifacial axis,** tend to be continuous or at least parallel; while in man,

or in related primates, they tend to form an obtuse angle which is progressively reduced from lower to higher types.

THE LIMB SKELETON.

It will be evident from a study of the limb skeleton of the rabbit that there is a general correspondence in structure as between its anterior and posterior divisions. This applies not only to the distinction of girdle portions, and the divisibility of the free extremity into proximal, middle and distal portions, but extends to very many smaller details of composition. The relation in structure as between anterior and posterior limbs is described as **serial homology,** since two structures in the same animal cannot be homologous in the usual meaning of the term.

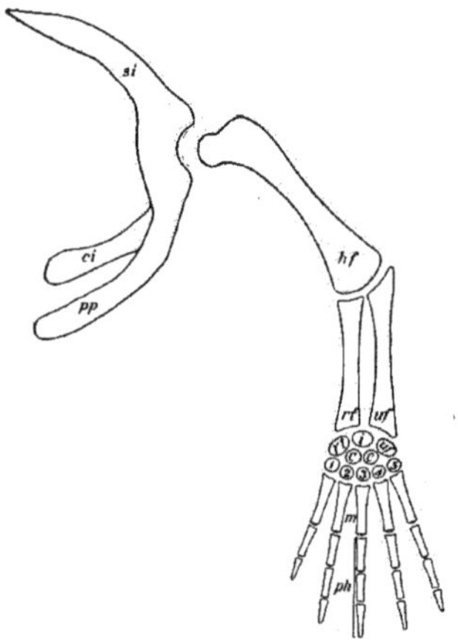

FIG. 32. Plan of the anterior limb skeleton in walking vertebrates, the equivalent elements of the posterior limb indicated in brackets: si, scapula (ilium); pp, procoracoid (pubis); ci, coracoid (ischium); hf, humerus (femur); rt′, radius (tibia); uf′, ulna (fibula); rt, radial carpal (tibial tarsal); i, intermedium; uf, ulnar carpal (fibular tarsal); c,c, centrals; 1-5, distal carpals (tarsals); m. metacarpals (metatarsals); ph, phalanges of the digits.

The pelvic girdle is more uniformly and solidly developed than the pectoral girdle, as shown by its strength in the three principal directions about the point of attachment of the limb, the great development of the ventral union, and strong attachment to the sacrum. On the other hand, the pectoral girdle is notable chiefly for the strong development of the scapula as opposed to the weakness of the ventral, pectoral portion. In the rabbit the clavicle is at best vestigial, and in many mammals, such as the ungulates, it is entirely absent. In man, however, as well as in other mammals, many of them primitive, in which the pectoral muscles of adduction are well developed, together with the corresponding movement of shoulder and arm, the clavicle is both well developed and articulated at the two ends with the scapula and sternum.

Both pectoral and pelvic girdles in terrestrial vertebrates conform more or less closely to a tri-radiate shape if the two halves

are considered individually. The corresponding portions may be identified (Fig. 32), though in mammals the ventral portion in the case of the pectoral girdle is greatly reduced. Of the ventral elements, those commonly present in vertebrates are the coracoid, which forms the posterior ray, and the **procoracoid**, the anterior ray, the latter being partly covered in front by a derm splint, the clavicle. This condition though not characteristic of mammals generally is still found in monotremes, and rudiments of the coracoid extension ventrally are

GROUND PATTERN IN TERRESTRIAL VERTEBRATES.

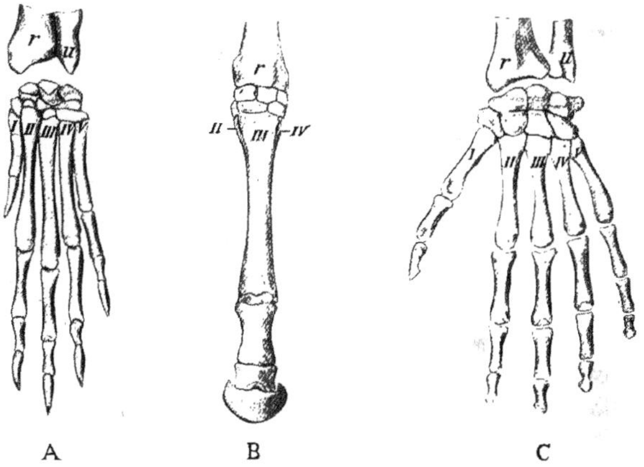

FIG. 33. Homologies of the mammalian limb. A, fore foot, rabbit B, fore foot, horse. C, human hand. r, radius; u, ulna; I-V, metacarpa bones.

identifiable in embryonic marsupials. Adult marsupials and placentals show only a small hook-like coracoid process, together with the clavicle in perfect or less perfect development.

There is no more striking feature of homology than that shown by the free extremities in respect of the different forms of vertebrates. This is **true homology** because it concerns the resemblances, part for part, in the anterior or in the posterior limb of any one vertebrate as compared with the corresponding elements in the same position in other forms. The front limb of the rabbit (Fig. 33) is slightly elongated, and semi-digitigrade, making it a more efficient

organ for running, though not so highly specialized as that of the horse. The human hand retains a fairly primitive form as to its general proportions, but is modified into a seizing or grasping type, the thumb being opposable to the remaining digits. All three kinds of limbs are, however, in themselves, modifications of a primitive, five-toed limb, sometimes termed the ideal **pentadactyl plantigrade** type, in which the palm of the hand or sole of the foot is placed flat on the ground. The composition of this primitive limb, traceable in one form or another throughout the higher vertebrate; and also the serial homologies of the parts are indicated in Fig. 32.

THE MUSCULAR SYSTEM.

The contractile tissues are not arranged in a definite continuous system as is the case with most other organ complexes of the body. Smooth or **involuntary muscle** fibres, modified mesenchyme cells of the embryo, which are under the control of the sympathetic nervous system, form the muscle coats of the digestive tube, and are important not only for its repeated, peristaltic movements, but also for its elasticity and expansive power. Smooth muscle is also a constituent of many other visceral organs, especially glands, in the active secretion of which it appears to play a mechanical part. It is further distributed through the walls of the bloodvessell, especially the arteries, where it forms the mechanical organ of the **vasomotor function.** This consists in the control of the diameter of the vessels by **vasodilator** and **vasoconstrictor nerves** connected with the vague nerve and the sympathetic nervous system. The constrictive action is stimulated by secretion of the suprarenal glands. Vessel regulation of this kind is important, first, in maintaining tone and therefore pressure, and, second, in controlling heat loss from the surface of the body. Action of the vasomotor nerves may be demonstrated physiologically in a variety of ways. Transection of the cervical sympathetic nerve of one side in the living rabbit is followed by vasodilation of the ear, the congestion of which can be seen, and the heat loss is demonstrable by feeling with the hand. Stimulation of the cephalic cut end is followed by vasoconstriction.

INVOLUNTARY MUSCLE.

Cardiac muscle, most nearly allied in action to smooth muscle is the mechanical organ of the rhythmical contraction or **beat** of the heart. The contraction takes place according to the succession of the chambers or the course of the blood, and the rate and strength of the beat is under the control of the vagus and sympathetic nerves, the action of which can be experimentally demonstrated. The excised heart in the case of lower vertebrates continues to beat for some time automatically or under stimulation. This behaviour has been interpreted as purely automàtic action of the heart muscle, but may depend upon intercardiac nerve connections.

THE HEART.

The **voluntary muscles** of the body form the nearest approach to a continuous system of all contractile tissues. They consist for the most part of parallel fibres, the association of which into fasciculi is responsible for the appearance of longitudinal striping when the gross muscle is viewed from the side and more or less for the grained appearance of the cut surface. The attachments, usually of white fibrous connective tissue occurs as concentrated **tendons** or as flat thin **aponeuroses.** The control of action is exercised directly from the spinal cord.

VOLUNTARY MUSCLE.

A typical muscle of the skeleton has the disposition of parts illustrated in Fig. 34 by the biceps (a flexor of the forearm) and the long head of the triceps (an extensor of the forearm). The fixed tendon, or **tendon of origin,** of the biceps is in relation to the glenoid border of the scapula, the movable tendon, or **tendon of insertion,** with the lower border of of the ulna. Noting the position of the muscle in front of the elbow joint, it will be seen that its contraction results in flexion, i.e., in bringing the forearm into a position nearer the arm, or of raising the forearm and hand from the ground. The analogous action of the triceps is similarly demonstrated, and it will also be evident that the immediate result of contraction upon the bones is limited by the form of the joint. In this case a hinge-joint confines motion to one plane, while in the case of the shoulder and hip, a joint of the ball-and-socket type allows motion on points at various angles to a plane according to which muscle or group of muscles may be brought into action.

A TYPICAL MUSCLE.

Voluntary muscle arises chiefly from the segmented areas or **myotomes** of the embryo. The extent to which segmentation is shown in the adult, however, depends for the most part on how far the definitive muscle is removed from the vertebral column or segmented portions of the skeleton. The vertebral muscles themselves show throughout their attachments to successive vertebrae the marks of segmental origin, but many others, such as those of the abdomen, to a certain extent those of the limbs, and those of the eye show practically no indications of their segmental origin.

ORIGIN.

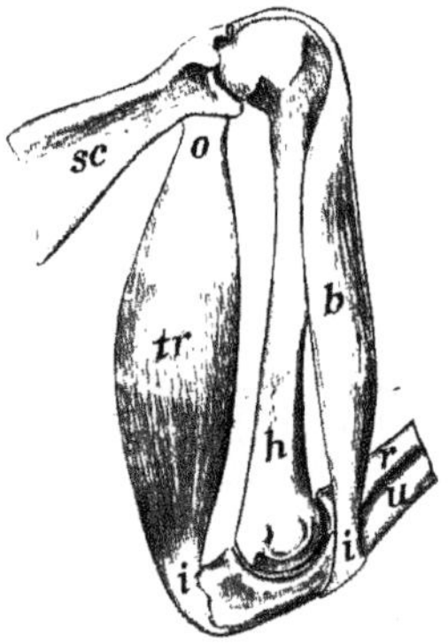

FIG. 34. Arm muscles of rabbit from the medial surface, illustrating muscle action, flexion and extension. b, biceps (flexor); tr, long head of triceps (extensor); i, insertion; o, origin; sc, scapula; h, humerus; r, radius; u, ulna.

While the bulk of voluntary muscle is **skeletal**, part at least is related to the skin. This forms a **cutaneous** sheet, divisible into the cutaneus maximus and platysma, commonly developed to a considerable extent in mammals, but reduced in man, and into the facial, palpebral, and auricular muscles of the head. The trunk musculature comprises a special group of **cervical** and **occipital** muscles in relation to the neck and head, and the general series which are more nearly **vertebral.** The **appendicular** muscles are distributed in special groups connecting the limb as a whole with the trunk and the various segments of the limb with one another.

DISTRIBUTION.

As already indicated the skeletons of anterior and posterior limbs are considered to be equivalent part for part. This is also true of the related muscles, but the respective positions of the segments of the limb in a mammal are such that in examining the muscles, it is important to determine the equivalence of the bone surfaces. Since also certain conventional ideas of flexion and extension prevail in human anatomy, it is equally important to establish the corresponding relation as between man and a quadrupedal form. The homologous surfaces may be

EQUIVALENCE OF LIMB MUSCLES.

determined by a study of the embryonic relations of the limbs, but may be referred more easily, if not so accurately, to the neutral plan as illustrated in Fig. 35.

The neutral type (A) may be defined as one in which the proximal segment (humerus or femur) is situated at right angles to the median vertical plane, the middle segment directed downward parallel to the median plane, and the distal segment again at right angles to it. In this condition the bones of the middle segment are parallel, with the radius, or the tibia, according to the particular limb, anterior in position. The first digit is also anterior. The entire anterior surface is indicated in the accompanying diagram by shaded lines. The angles b and c are "extension"

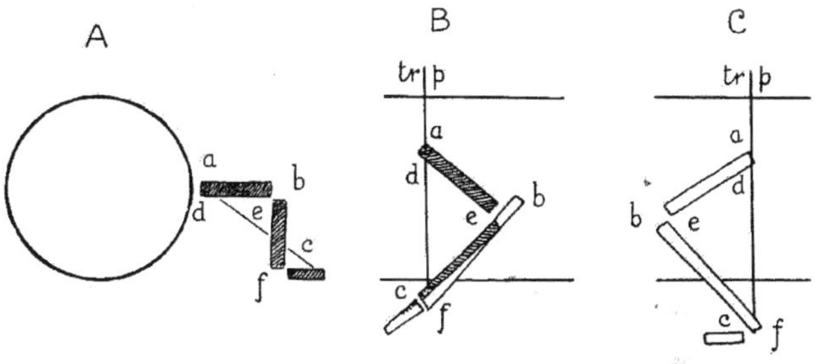

FIG. 35. Schematic representation of the respective positions of the segments in the mammalian limbs: A, neutral; B, anterior limb; C, posterior limb. Explanation in text: tr.p., transverse plane.

angles; e and f "flexion" angles; a is an "abduction" angle, while d is an "adduction" angle.

In both limbs of a mammal, the entire appendage is rotated downward to a position more or less underneath the body. In the anterior limb (B) the proximal segment is rotated backward, the middle and distal segments forward. The two divisions thus come to differ from one another to the extent of two right angles. Also, in the front limb, the interesting condition is observable that the radius and ulna are crossed on one another, the position of the former being chiefly anterior, although its proximal end is lateral, and its distal end, like the first digit, medial. In the rabbit, as in most mammals, the hand is thus fixed in a prone position, while in

man the hand may be rotated to a supine position in which the radius is wholly lateral, or in which the two bones are parallel.

In the posterior limb (C) all three segments are rotated forward. Consequently the bones of the middle segment retain their original parallel position with reference to one another. The extension angles of the knee and ankle are anterior, the flexion angles posterior. They differ from the corresponding angles of the anterior limb only at the middle joint, but here the difference amounts to two right angles. It will be observed also that the distal or ankle joint of the posterior limb retains a primitive condition, in which the foot is placed approximately at right angles to the leg; in other words, it is of a plantigrade type, one in which the sole of the foot rests on the ground. Thus the muscles described as flexors in the foot pass over the heel. They are functionally flexors of the toes, but extensors of the foot as a whole. Those described as extensors lie for the most part on the anterior surface of the tibia, and are turned from their original course to the dorsal surface of the foot. They are functionally extensors of the toes, but flexors of the foot as a whole.

THE NERVOUS SYSTEM.

There is probably no system of organs in which external form is so little suggestive of actual function as is the case with the nervous system. This is perhaps less true of its peripheral portion, consisting of nerves which can be seen ramifying through all parts of the body, than of the central portion comprising the brain and spinal cord. In examining the external form as a preliminary step to the study of the functional arrangements, it is advisable to bear in mind that the nervous system is a great correlating mechanism, consisting of central **cellular portions**, and of **conducting paths** to and from these centres to outlying parts of the body.

In accordance with its prime importance and at the same time the non-resistant character of the tissue of which it is composed, the central nervous system is protected within the canal of the vertebral column and cavity of the brain cases. It is furthermore surrounded by connective tissue membranes of which one, the

dura mater, forms a tough external investment, while another, the **pia mater**, is a more delicate, highly vascular membrane lying next the nervous matter, and amply supplied with blood-vessels for the supply of nourishment. The nerves, however, are distributed freely throughout the body, and though not so adequately protected are more capable of withstanding mechanical stress, and are commonly found in connective tissue situations where mechanical injury is not likely to occur.

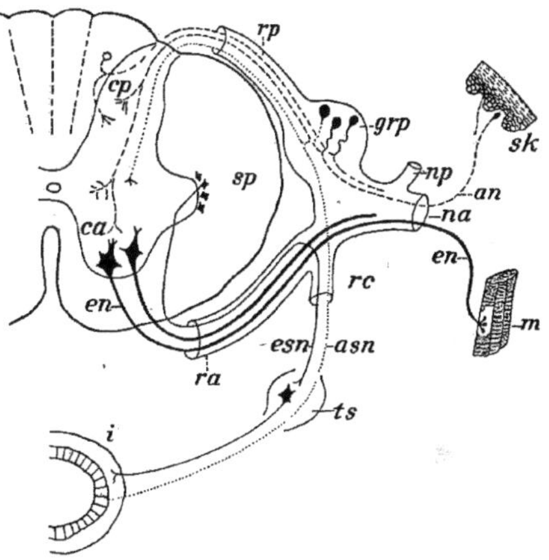

FIG. 36. Plan of the central and peripheral connections of a spinal nerve: an, afferent (sensory) neurone; asn, afferent sympathetic neurone; ca, cp, anterior (ventral) and posterior (dorsal) columns of grey matter; en, efferent (motor) neurone; esn, efferent sympathetic neurone; grp, dorsal root ganglion; i, intestine (visceral organs); m, skeletal muscle; na, np, anterior (ventral) and posterior (dorsal) rami of spinal nerve; ra, rp, anterior and posterior roots of spinal nerve; rc, ramus communicans (sympathetic); sk, skin; sp, white matter of cord; ts, ganglion of sympathetic trunk. Slightly modified, from Herrick.

The most typical of the structural arrangements of the nervous system may be made out from a study of the connections of any one of the paired nerves of the spinal series (Fig. 36). In the spinal cord the difference in appearance as between the white and the grey matter has already been described (p. 39). A spinal nerve arises by two **roots**, one of which is **dorsal** and bears a small **ganglion** of cellular material, the other **ventral** and without a

ganglion. Impulses passing through the dorsal root are **centripetal** or afferent in that they pass only in the direction of the central nervous system, but they are also in many cases **sensory** in that their effects may be consciously experienced. The most characteristic sensory impulses are those which come from the skin. In a similar fashion the impulses of the ventral root are **centrifugal** or **efferent,** in that they pass only in a direction away from the central nervous system, and are in most cases **motor** in that their effects are commonly observed as muscular contraction. The two roots, however, unite immediately outside the spinal cord, and subsequently re-divide in such a way that three nerves are formed, each containing a proportion of both kinds of fibres. Two of these nerves, known as the **dorsal** and **ventral rami** and then distributed as somatic nerves to the body wall, while a third is distributed as a visceral nerve, or **ramus communicans,** having important secondary connections in the sympathetic nervous system. Unlike the somatic nerves, which take a direct course to their terminations, the communicating rami of each side, unite in a position ventral to the vertebral column to form a longitudinal **sympathetic trunk** consisting of a connected series of **ganglia.** The latter is similarly connected with a **prevertebral** series of the ganglia, and through them with certain **peripheral** ganglia on the surface of the visceral organs. This system forms the sympathetic division of the peripheral nervous system. Its ganglia are nerve-cell centres, and its fibres, afferent and efferent, are distributed both to the visceral organs and to bloodvessels in all parts of the body.

COMPOSITION OF A SPINAL NERVE.

It is difficult to determine what portion of a muscular contraction, even if considered to be purely voluntary, actually arises from an impulse originating in the central nervous system. The animal body, however, affords many indications of muscular actions as **responses** to previous incoming **stimuli,** without conscious experience as a necessary factor in the result. Stimulation of the skin, in vertebrates in which the spinal cord is divided, and thus separated from the brain, is followed by muscular contractions, presumably by direct connections of individual dorsal and ventral roots, or extension of the stimulus to neighbouring roots. This is known as **reflex action.** Its nature and conditions can be determined by

REFLEX ACTION.

using spinal or decerebrated frogs, but similar actions, of which the well-known scratch reaction of the dog is an example, are common to all animals.

THE SPINAL CORD.

The spinal cord reflects in its form the ground type of the central nervous system. Developed in the embryo as a tube, it retains this condition throughout life. The cavity, however, is reduced to a slender **central canal**, lined by a thin epithelial membrane, the **ependyma**, while through proliferation of its cells, and their fibre extensions, the walls become enormously thickened, and the nervous functions of the system, as a connected conducting mechanism, thereby established. The cord traverses the vertebral canal, showing slight

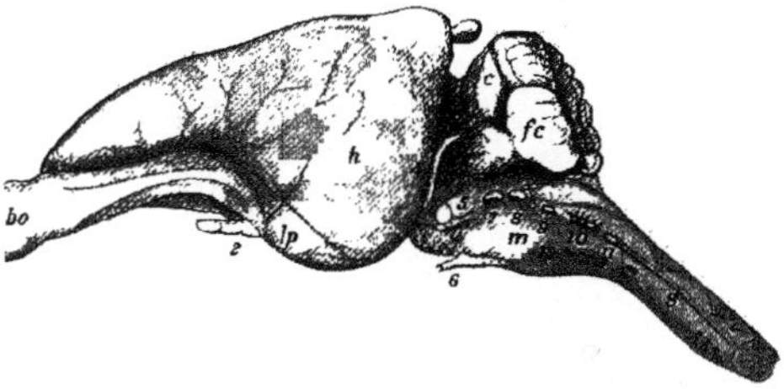

FIG. 37. The brain from the left side: bo, olfactory bulb; c, cerebellum; fc, paraflocculus cerebelli; h, cerebral hemisphere; lp, piriform lobe (olfactory brain); m, medulla oblongata; s, spinal cord. Numerals indicate the corresponding cranial nerves; 2, optic; 4, trochlear; 5, trigeminal; 6, abducens; 7, facial; 8, acoustic; 9, glossopharyngeal; 10, vagus; 11, spinal accessory; 12, hypoglossal.

enlargements in the cervical and lumbar regions in relation to the nerve supply of the limbs, and at about the level of the second sacral vertebra narrows into the slender, thread-like **filum terminale,** by which it is continued almost to the middle of the length of the tail.

Superficial examination of the brain of the rabbit (Fig. 37) shows that its larger part is formed by the paired **cerebral hemispheres.** They are closely pressed together on the dorsal side but separated posteriorly on the ventral side. The external layer, important as forming the cellular **cortex,** is smooth in the rabbit as in rodents generally, but is corrugated in higher mammals such as the cat and dog, while in man it reaches a high degree of elabora-

tion. The anterior tips of the cerebral hemispheres in the rabbit are slightly expanded to form the **olfactory bulbs**, and the connection of the latter with the posteroventral portion of the hemisphere known as the **piriform lobe** is easily traced. This portion of the brain is chiefly of interest in mammals because of its relation to the olfactory sense organs and because of the reduced condition of the organ in man. The development of the cerebral hemispheres even in the lower mammals is such that important parts of the brain, notably the diencephalon and mesencephalon, are dorsally and laterally concealed. On the ventral side, however, the **chiasma** or crossing of the optic nerves is evident, while more posteriorly the divergent cords of the **cerebral peduncles** may be seen passing forwards in the direction of the hemispheres. The posterior part o the brain is formed largely by the **cerebellum** above, the corrugation of which is one of its outstanding features, and by the somewhat tapering brain-stem by which the brain is connected with the spinal cord. Its bulk is formed by the **medulla oblongata**, which is crossed in front by a bridge of fibres, not so conspicuous in the rabbit as in many mammals and in man, which is known as the **pons** and which connects the two sides of the cerebellum. These, the outstanding surface features of the brain, afford but a moderate conception of its details, the nature of which can only be made out by more thorough examination and by reference to the plan of development of the organ as a whole.

THE BRAIN.

Like the spinal cord, with which it is continuous, the brain forms primarily a portion of the **neural tube**, containing a central cavity or **neurocoele**, but unlike the spinal cord it is greatly enlarged and elaborated both as the general centre of the whole nervous mechanism and also the special centre of the nervous mechanism for a variety of functions connected with the head. It accordingly forms not only a more or less distinct division, known as the brain or **encephalon**, as opposed to the more general division, the spinal cord or **spinal medulla**, but is also divided into a series of paired and unpaired segments, containing divisions of the original cavity in the form of **ventricles**.

The primary divisions of the brain are more or less similar and homologous in all vertebrates. The more elaborate condition of the

organ in a mammal may be explained by reference to the general plan as indicated in Fig. 39, which is based upon general features of form in vertebrates and upon embryonic development. For comparison in the gross the brain of the frog (Fig. 38) offers one of the best examples.

The brain as first formed in the embryo appears as an anterior expanded portion of the neural tube, or rather as three expansions arranged in a linear series. They are described as the **primary cerebral vesicles;** or, as primary divisions of the future brain, they are designated in anteroposterior order as the **prosencephalon, mesencephalon,** and **rhombencephalon.**

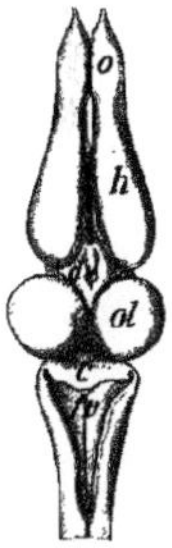

FIG. 38. The brain of the frog from the dorsal surface. c, cerebellum; d, diencephalon; fv, fourth ventricle; h, cerebral hemisphere o, olfactory lobe; ol, optic lobe.

The first of the primary divisions, the **prosencephalon,** or **primary fore-brain,** becomes divided during development into two portions, namely, an anterior portion, the **end-brain** or **telencephalon,** which is largely a paired structure, and a second portion, unpaired, the **diencephalon,** or **inter-brain.** The larger paired portion of the telencephalon is the basis of the **cerebral hemispheres.** It contains, as divisions of the primary cavity, a pair of cavities, the **lateral ventricles.** The anterior portion of the telencephalon, moreover, becomes differentiated, so that a small terminal olfactory segment, the **rhinencephalon,** is more or less perfectly marked off from the rest. In the mammalian brain this part is chiefly identifiable as the paired **olfactory bulb,** the latter being the anterior portion of the olfactory lobe or **olfactory brain,** and containing in its interior an extension of the lateral ventricle.

The unpaired portion of the prosencephalon is considered as belonging in part to the telencephalon and in part to the diencephalon. Its cavity, the **third ventricle,** is connected with the lateral ventricles through the **interventricular foramen.** Its anterior wall is formed by a transverse connection of the cerebral hemispheres, the **lamina terminalis.** In all vertebrates this portion of the brain is remarkable for the manner in which its wall is differentiated. The ventral portion extends downward as a

slender funnel-like structure, the **infundibulum,** the tip of the latter being in contact with the **pituitary body** or **hypophysis** and its base connected with a small grey elevation, the **tuber cinereum.**

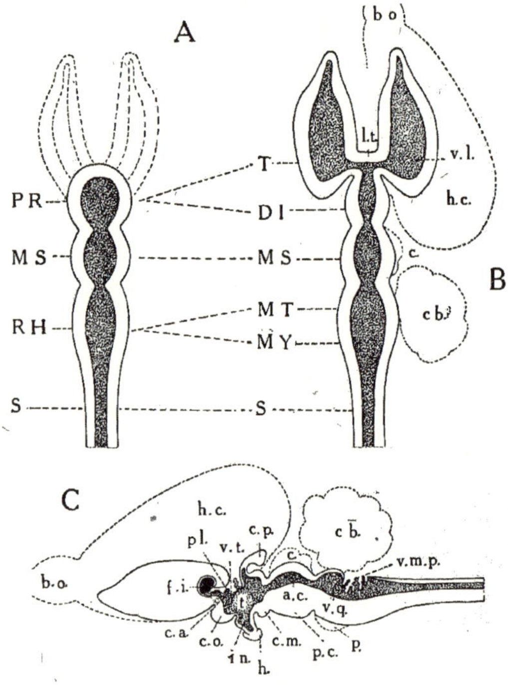

FIG. 39. Plan of the divisions of the vertebrate brain: A, embryonic; B, adult, projection from dorsal surface; C, adult, sagittal section. The contour of the mammalian brain is indicated by broken lines.
Primary divisions—PR, prosencephalon; T, telencephalon; DI, diencephalon; MS, mesencephalon; RH, rhombencephalon; MT, metencephalon; MY, myelencephalon; S, spinal cord.
a.c., cerebral aqueduct; b.o., olfactory bulb; c., corpora quadrigemina; cb., cerebellum; c.m., mamillary body; c.o., optic chiasma; c.p., pineal body; f.i., interventricular foramen; h., hypophysis; h.c., cerebral hemisphere; in., infundibulum; l.t., lamina terminalis; p., pons; pl., chorioid plexus of third ventricle; p.c., cerebral peduncle; t., thalamus, also indicates position of massa intermedia; v.l., lateral ventricle; v.m.p., posterior medullary velum; v.q., fourth ventricle.

Its cavity is the **recessus infundibuli.** Immediately in front of the infundibulum the ventral portions of the optic tracts join to form the **optic chiasma,** and immediately behind it the floor is

thickened, forming externally a pair of rounded protuberances, the **mamillary** bodies. In the brain of the rabbit this structure consists superficially of a larger median portion with faint lateral elevations appended to it. Collectively, these structures are considered to form a major division, the **hypothalamus**, the latter consisting of two portions, namely, an **optic portion**, comprising the infundibulum, tuber cinereum, and the optic chiasma, and a **mamillary portion**, including the mamillary bodies. The two portions are commonly classified as belonging respectively to the telencephalon and the diencephalon, but embryological study places the boundary of these of the latter divisions at the optic recess or roughly at the point of the optic chiasma.

The more dorsal portion of the diencephalon, containing the major part of the third ventricle, is known as the **thalamencephalon.** Its lateral walls are greatly thickened, while its roof is extremely thin, especially in its anterior part. Here the actual roof of the ventricle is formed only of a thin layer of tissue, the **epithelial chorioid lamina,** but the latter has associated with it a series of vascular ingrowths of the investing pia mater, the latter being described in this relation as the **chorioid web** (tela chorioidea). The two structures together form a **chorioid plexus.** This extends downward into the third ventricle, reaching out also into the lateral ventricles.

The dorsal portion of the thalamencephalon bears posteriorly the **pineal body,** the latter together with certain related structures, the **habenulae** and **habenular commissure,** forming the **epithalamus.** The general portion of the thalamencephalon bordering the third ventricle, and broadly connected across the latter by the **massa intermedia,** is the **thalamus.** In the brain of the rabbit it will be seen that the thalamus is chiefly indicated externally by a rounded protuberance, the **pulvinar.** The latter is dorsal in position and is imperfectly marked off from a second protuberance, the **lateral geniculate body,** lying on its postero-lateral side. To the medial side of this is a third protuberance, the **medial geniculate body.** The medial and lateral geniculate bodies as thus defined constitute the **metathalamus** (Fig. 84).

The second of the primary divisions, the **mesencephalon,** or **mid-brain,** is noteworthy in a mammal as lacking a ventricle.

Its cavity is a narrow canal, the cerebral **aqueduct,** leading from the third ventricle backward to the fourth ventricle, or cavity of the rhombencephalon. Externally, its roof is differentiated into four rounded elevations, the **corpora quadrigemina,** of which the anterior pair are much larger than the posterior ones. Its floor is chiefly formed by a pair of divergent cords, the **cerebral peduncles.**

The parts of the mesencephalon and prosencephalon together constitute the large brain, or **cerebrum.**

The third primary division, the **rhombencephalon**, or primary **hind-brain**, is a greatly elaborated portion from which arise the majority of the cranial nerves. The constricted area joining it with the mesencephalon is known as the **isthmus rhombencephali.** It includes the **anterior medullary velum** and **brachia conjunctiva** (Fig. 86). The rhombencephalon itself is divisible into two portions, especially well defined in the mamalia, namely, the **metencephalon,** or **hind-brain,** and the **myelencephalon,** or **after-brain.** The former includes the **small brain,** or **cerebellum,** and a ventral structure of a commissural nature, the **pons.** The myelencephalon is a transitional portion connecting the brain with the spinal cord. The cavity of the rhombencephalon is the **fourth ventricle.** It is a peculiarly shaped space, the floor and lateral walls of which are very greatly thickened, while the roof is for the most part thin. The roof appears at first sight to be formed largely by the cerebellum, but is in reality formed by two membranes underlying the latter. One of these, the **anterior medullary velum,** is connected forwards with the mesencephalon, while the other, the **posterior medullary velum,** covers a triangular space over which the posterior margin of the cerebellum does not extend. The posterior medullary velum has the same structure as the chorioid plexus of the third ventricle, but is much more poorly developed.

Apart from its principal divisions, which, as indicated above, are more or less common to all vertebrates, the brain depends for its external form on the elaboration of certain parts in comparison with others. In the mammalia the cerebral hemispheres and the cerebellum are the chief form-determinants, although the pons and the corpora quadrigemina may also be considered in this connection.

PLAN OF THE DIVISIONS OF THE BRAIN

(ENCEPHALON)

I. PROSENCEPHALON (Anterior cerebral vesicle) (Primary fore-brain)	Telencephalon (End-brain)	*Cerebral hemisphere*	*Olfactory lobe* (rhinencephalon)	**Cerebrum**
			Corpus striatum, pallium and minor divisions	
		Optic portion of hypothalamus	*Hypothalamus*	
	Diencephalon (Inter-brain)	*Mamillary portion* of hypothalamus		
		Thalamencephalon	*Thalamus*	
			Metathalamus	
			Epithalamus	
II. MESENCEPHALON (Middle cerebral vesicle) (Primary mid-brain)	- - - - - - - - - - -		*Corpora quadrigemina*	
			Cerebral peduncles	
III. Rhombencephalon (Posterior cerebral vesicle) (Primary hind-brain)	Metencephalon (Hind-brain)	Cerebellum		
		Pons		
	Myelencephalon (After-brain, or medulla oblongata)			

It will be seen also that the form of the brain is more or less dependent on the existence at certain places of well-marked flexures (cf. Plate II). The first of these, the **cephalic flexure,** is in the region of the mesencephalon, the anterior portion of the brain being bent downward; the second, or **pontine flexure,** is at the fourth ventricle; while the third, or **cervical flexure,** is at the point where the myelencephalon passes over into the spinal cord.

The peripheral nervous system embraces two groups of paired and, for the most part, metamerically arranged nerves, namely, the **spinal nerves**—those arising from the spinal cord and leaving the vertebral column through the intervertebral foramina; and the **cranial** or **cerebral nerves**—those arising from the brain and passing through the foramina of the skull. Of these the spinal nerves (p. 68) are less modified, both in structure and distribution.

The **cranial nerves,** those arising from the brain, and making their exit through the walls of the skull, are comparable in some respects to the spinal nerves, but in many ways are different in nature in addition to being in some cases highly specialized. Three pairs, respectively, **olfactory, optic,** and **acoustic,** or **first, second,** and **eighth** of the series are afferent nerves from the special sense organs of smell, sight and hearing, the function of the acoustic nerve including also maintenance of equilibrium. The **third, fourth** and **sixth** nerves, respectively, **oculomotor, trochlear** and **abducent,** are distributed as somatic motor nerves to the muscles of the eyeball, but also contain fibres of muscle sense.

THE CRANIAL NERVES.

Of the remaining cranial nerves the fifth, seventh, ninth and tenth are branchiomeric. Although the connections of these nerves are not fully considered in the dissection as here outlined, their chief characteristic as branchiomeric structures may be indicated. The **fifth,** or **trigeminal** nerve, the nerve of the mandibular arch, arises in two parts, one of which, the portio major, is sensory, the other, the portio minor, motor. The portio minor unites with the third or mandibular division of the portio major. Thus, the terminal branches of all three divisions, ophthalmic, maxillary, mandibular, are distributed as somatic sensory nerves to the skin of the head, and, in addition, the mandibular nerve

distributes visceral motor branches to certain muscles (masticatory group, mylohyoid and digastric) regarded as belonging to this, the first arch. A visceral sensory connection with the mouth is considered to be formed by the lingual branch of the mandibular nerve and by the palatine branches of the spheno-palatine ganglion. Both are, however, connected with the central nervous system through the seventh nerve, the former by the chorda tympani, and the latter by the great superficial petrosal.

The **seventh,** or **facial nerve,** is the nerve of the second, or hyoid arch. It is chiefly distributed as a visceral motor nerve to the cutaneous muscles of the head, but contains also taste fibres from the tongue. The **ninth,** or **glossopharyngeal** nerve, belonging to the third arch, the **tenth,** or **vagus,** belonging to the fourth and succeeding arches in lower forms, and the **eleventh,** or **spinal accessory** nerves, the latter apparently related to the vagus as a motor portion, are distributed as visceral motor nerves to the pharyngeal and laryngeal musculature, and as visceral sensory nerves to various visceral organs, the ninth nerve supplying the gustatory organs of the tongue. The vagus contains a variety of fibres, both afferent and efferent, the former from the larynx and respiratory organs, the latter distributed to the organs of circulation and digestion. The spinal accessory has a characteristic distribution to the cleidomastoid, sternomastoid and trapezius muscles of the side of the neck and shoulder. The **twelfth,** or **hypoglossal** nerve has the relation of the ventral or motor portion of a spinal nerve, and is distributed as a motor nerve to the muscles of the tongue.

THE DIGESTIVE SYSTEM.

The **digestive system** comprises as its chief portions the **digestive tube** and the **digestive glands.** The digestive tube is divisible into several parts, which, with the exception of the caecum and its vermiform process, are arranged in a linear series. The digestive glands comprise the **oral glands,** the **liver,** and the **pancreas.** They are parts of an extensive series of epithelial glands, otherwise contained within the wall of the tube, and for this reason not appearing as gross structures.

The parts of the digestive tube may be classified as follows:

1. ORAL CAVITY.
   - Oral cavity proper.
   - Vestibulum oris.

2. PHARYNX.
   - Nasal portion.
   - Oral portion.
   - Laryngeal portion.

3. OESOPHAGUS.

4. STOMACH.

5. SMALL INTESTINE.
   - Duodenum.
   - Mesenterial intestine.
     - Jejunum.
     - Ileum.

6. LARGE INTESTINE.
   - Caecum.
   - Vermiform process.
   - Colon.
   - Rectum.

DIGESTION AS A PROCESS.

The digestive system comprises a variety of functions, both mechanical and chemical, and connected directly and indirectly with the digestion of food. In the oral cavity solid food is comminuted by the action of the teeth, and is mixed with salivary secretion, so that it is more easily swallowed and passed along the œsophagus to the stomach. The secretion of the oral glands is thus important chiefly for its mucous element, but that of the parotid especially contains an enzyme, **ptyalin**, which is capable of converting starch into soluble material. Food is further reduced to a pulp-like mass in the stomach, while the gastric secretion, containing **pepsin** and **rennin**, exercises a dissolving action upon proteid, and a coagulating action upon milk. The liver secretion, known as **bile**, contains, in addition to coloring materials, salts which exert a splitting action upon fats. The pancreatic secretion contains a variety of enzymes, converting proteids and starches, and breaking fats into fatty acids and glycerin. The actions of the dissolving enzymes is successive, secretion being dependent to some extent on antecedent bodies by which the stimulus for secretion is determined. The preliminary processes of digestion refer in this way to the mechanical action of food passage along the canal, and to the provision of converting enzymes. Absorption, which the final object of the digestive process is accomplished chiefly in the large intestine through the bloodvessels and lymphatics of

the wall. The relatively great extent of the wall, including the enormous development of the caecum in the rabbit and other rodents, is related to the comparatively great bulk and low nutritive quality of the ingested food.

In its most general features the digestive system is significant as an epithelial tube, in which the food is modified, by solution or otherwise, so that it is capable of being absorbed through the epithelial surface. In the form of the digestive tube as seen in a vertebrate, however, a number of gross mechanical features are evident, such, as, for example, the increase in capacity, or in absorptive area, through the folding of the mucous membrane, or the expansion of the wall; or again, the presence of a special muscular tunic, and its modification at certain places, as in the

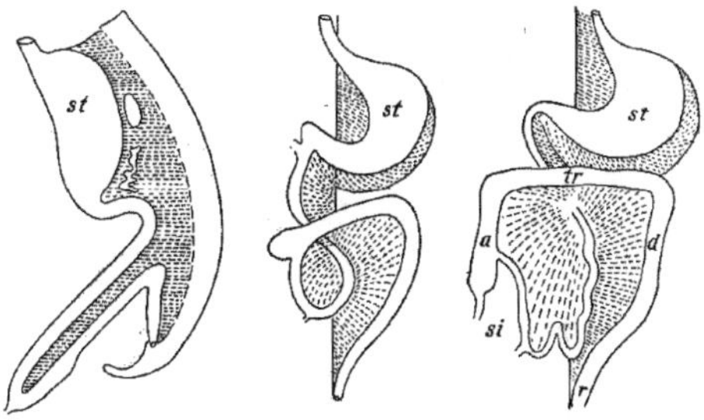

FIG. 40. Plan of successive embryonic stagesin displacement of the digestive tube and common mesentery from the midline position (man): a, tr, d, ascending, transverse, and descending colons; r, rectum; si, small intestine; st, stomach Modified from figures by Toldt and Hertwig.

oesophagus, the pyloric limb of the stomach, and the first portion of the colon. Moreover, many features of the abdominal portion of the tube, and, indeed, certain of its recognized divisions, depend on its relation to an extensive serous sac—in a mammal the peritoneal cavity. In this connection it is to be considered that the digestive tube is primarily a median structure. It has this relation in the earlier stages of embryonic development (Figs. 23, 40), and in many of the lower vertebrates it does not deviate to a great extent from a median position. In all higher vertebrates, however, the tube becomes greatly elongated in comparison with the cavity in which it lies, and

FORM AND SYMMETRY.

thus becomes extensively displaced to one side or other of the median plane. This development, while advanced in all mammals, may be said to reach an extreme in the herbivorous mammalia; and in many cases it is further increased by the independent elaboration of the blind intestine or caecum. In the rabbit the combined length of the small and large intestines is approximately eleven times that of the body.

In considering the divisions of the digestive tube in the rabbit, the posterior, or post-cephalic portion, comprising the oesophagus and succeeding parts, may be distinguished from the anterior, or cephalic portion, the latter comprising the oral cavity and pharynx. The former is a free portion embracing the digestive tube proper, while the latter is a fixed portion exhibiting a variety of general mammalian features connected with the organization of the head.

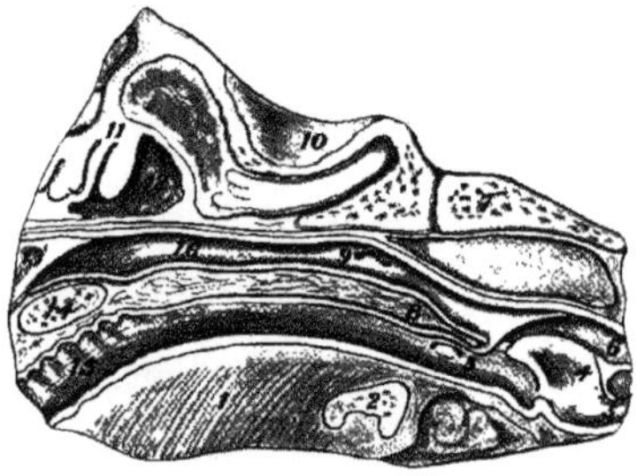

FIG. 41. The nasopharynx and related parts of the head as seen in median section: 1, tongue; 2, hyoid; 3, tonsil; 4, epiglottis; 5, entrance to trachea; 6, entrance to œsophagus; 7, basioccipital bone; 8, soft palate; 9, pharnygeal aperture of auditory (Eustachian) tube; 10, cranial cavity; 11, ethmoturbinal scrolls; 12, nasal cavity; 13, nasal septum; 14, hard palate; 15, oral cavity.

The form of the anterior, or cephalic portion of the digestive tube (Plate II) depends on its fixed relation with respect to the enclosing parts of the head-skeleton. In the rabbit, as in mammals generally, the **oral cavity** is divisible into two portions, of which one is the oral cavity proper, while the other, the **vestibulum oris**, is a space enclosed between the alveolar process of the jaws and the teeth on the one hand and the cheeks and lips on the other.

PRINCIPAL DIVISIONS.

As in other vertebrates, the **tongue** is a muscular structure projecting upward and forward into the oral cavity from its base of attachment on the hyoid apparatus, but its greater elaboration, as well as the differentiation of special processes, the circumvallate and foliate papillae, for the accommodation of the gustatory organs, are features of mammalian significance. The roof of the oral cavity is formed by an extensive palatal surface,

comprising the **hard palate,** and the **membranous,** or **soft palate.** These structures also form the floor of the accessory respiratory tracts of the nose, the posterior aperture being thus carried backward to a point more directly above the aperture of the larynx.

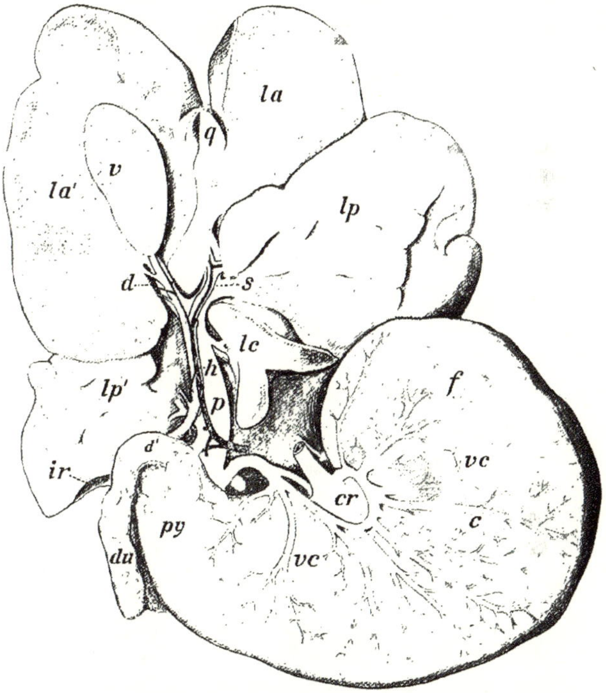

FIG. 42. The stomach, liver, and portal connections. Stomach: c, body; cr, cardia; f, fundus; py, pyloric limb; du, duodenum. Liver: ir, renal impression; la, lp, anterior and posterior lobules of the left lobe; la′, lp′, anterior and posterior lobules of right lobe; lc, caudate lobe; v, gall bladder. Vessels: d′, opening of common bile duct (shown as transparency); h, hepatic artery; p, portal vein; vc, tributaries of coronary vein. The ramifications of the portal vessels are indicated right (d) and left (s).

The chief features of the **pharynx** depend on its relation as a common or general portion of the digestive tube with the tubes of the respiratory system. It is divisible into an **oral portion,** representing the direct connection of the oral cavity with the oesophagus, a dorsal or **nasal portion,** connected with the nasal

fossae, and also with the middle ear through the internal auditory tube, and a ventral or **laryngeal portion,** containing the aperture of the larynx (Fig. 41).

The **oesophagus** is a slender but greatly expansible tube leading from the pharynx to the stomach. In its passage backward it traverses the neck and the thorax, and in both regions occupies a median position. In the thorax (Plate VII) it will be observed that it lies between the heart and the dorsal aorta, thus exhibiting the original relation of the digestive tube to the aortic portion of the vascular system. The succeeding portions of the digestive tube are those associated with the peritoneal cavity, and with the exception of the terminal portion, the rectum, are displaced from a median position. Consequently, the divisions which are recognized are based partly on the differential characters of the wall, and partly on the position of structures more especially in relation to the supporting peritoneum. Thus, the chief features of the **stomach** (Fig. 42) depend on the expansion of the organ and the rotation of its pyloric end forward and to the right. In the intestinal tract as a whole the chief, although by no means most conspicuous feature of position, depends on the looping of the entire structure on itself, so that the terminal portion, chiefly the transverse colon, crosses the ventral surface of the duodenum and then turns backward on the dorsal surface of the mesenterial small intestine. The **duodenum** is sharply marked off from the mesenterial intestine as an extensive loop, containing the major part of the pancreas and its duct, and lying on the right side of the dorsal wall of the abdomen. The **common bile duct** enters its first portion immediately beyond the pylorus. The **mesenterial intestine** is a greatly convoluted portion, lying chiefly on the left side of the abdominal cavity, and loosely supported by the broad, frill-like mesentery. From the pylorus to the sacculus rotundus there is no abrupt change in the character of the wall, although the first portion of the mesenterial intestine, that designated as the **jejunum,** and the duodenum may be considered together as a more vascular portion with thicker walls in comparison with the second portion, the **ileum,** in which the wall is less vascular and more transparent.

The main portion of the large intestine, the **colon,** although greatly specialized, may be considered to consist as in man of

ascending, transverse, and descending parts, that is to say the **ascending colon** lies on the right side of the body and passes in a general way from its point of origin on the caecum forward to a point where it becomes flexed to the left as the **transverse colon**; the latter crosses the body and is flexed backward as the **descending colon.** In the rabbit, however, that portion definable as the

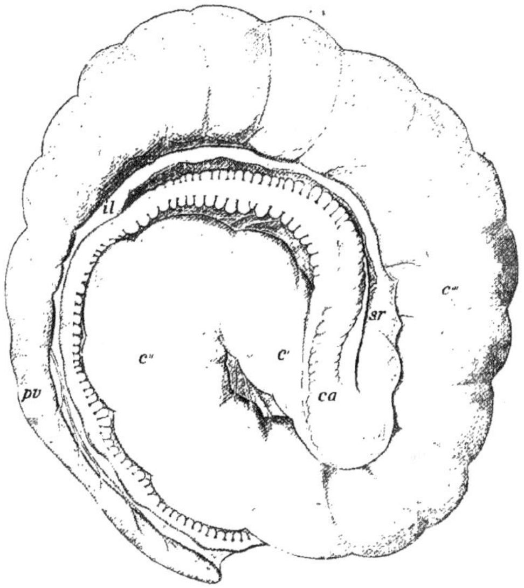

FIG. 43. The caecum and vermiform process: c', c'', c''', first, second, and third limbs of the caecum; ca, beginning of the ascending colon; il, ileum; pv, vermiform process (appendix); sr, sacculus rotundus.

ascending colon is greatly elongated, and is composed of five principal limbs, united by flexures. Two of these, in dissection from the ventral surface, are concealed by the base of the superior mesenteric artery, since they lie on its right side. The descending colon is also only nominally related to the left side of the body wall, its supporting peritoneum, the descending mesocolon, being closely connected with the mesoduodenum of the ascending limb of the duodenal loop. The course of the **caecum** (Fig. 43) as it lies in the body is comparable to two turns of a left-hand spiral, its blind

termination, the **vermiform process,** being dorsal in position and directed for the most part backward. It may be observed at this point that in their vascular supply the more typical divisions, namely, the transverse and descending colons, have arterial branches, respectively, the middle and left colic arteries, comparable to those of man; while on the other hand the right colic relation, on account

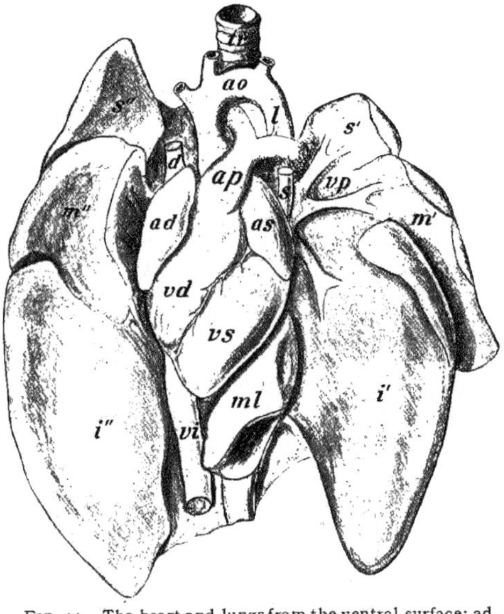

FIG. 44. The heart and lungs from the ventral surface: ad, right atrium; ao, aorta; ap, pulmonary artery; as, left atrium; d, right superior caval vein; i', i'', left and right inferior lobes of lung; l, aortic ligament; m', m'', middle lobes; ml, medial lobule of right inferior lobe; s', s'', superior lobes; tr, trachea; vd, right ventricle; vp, pulmonary veins; vs, left ventricle.

of the great complexity of its parts, is represented by a large number of vessels, branches of a common ileocaecocolic trunk.

THE RESPIRATORY SYSTEM.

In all air-breathing vertebrates the **lungs** (Fig. 44) are paired sacs which arise embryonically as ventral outgrowths of the digestive tube, and are secondarily connected with the outside of the body

through special perforations of the anterior portion of the head and through the oral cavity. The principal connection in a mammal is represented by an extensive nasal cavity bearing on its lateral walls the olfactory sense-organs. It is distinguished as an **accessory respiratory tract** from the true respiratory tract formed by the **trachea** and its terminal divisions, the **bronchi.** The respiratory system as represented by the lungs and related tubes, is nominally ventral to the oesophagus, but this relation is chiefly true of the trachea. In the thorax (Plate VII) the bronchi are, in general, interposed between the oesophagus and the heart, the lungs being expanded laterally into the paired pleural cavities.

In addition to carrying air over the sensory, olfactory surfaces, the respiratory system has accessory functions in relation to respiration. The mucous membrane of the nose, including that of the turbinated surfaces, serves both to warm the air, and to remove particles of foreign material. The chief function is, however, respiratory.

RESPIRATION AS A PROCESS.

In a mammal, respiration is both a physicochemical and a mechanical process. The former is fundamental, and consists in the supply of oxygen to the blood, and in this way to the tissues, for the oxidative phases of metabolism; also in the discharge of waste gases, principally carbon dioxide, from the blood to the air. The absorption and transport of oxygen is a specific function of the red blood cells. Though the oxygen, of which a certain amount always remains in the lungs during the process of breathing, must pass through the thin epithelial covering of the terminal air sacs into the capillaries before it can be taken into the blood cells, the latter from their flattened shape and very great numbers present a relatively enormous surface for absorption, the process being thereby facilitated. The lungs themselves are highly elastic, expansible sacs. They have the structure of greatly ramified saccular glands, except that the free internal surfaces are everywhere in contact with air. The division of the trachea into its bronchi, together with the bronchial ramifications, are the trunk and main branch portions of a rather complex system of tubes (Fig. 45), of which the terminal air-spaces are the final and functional parts.

What is commonly described as respiration, or the act of breathing, is a mechanical, muscular process accessory to respiration. It consists in the expansion of the thorax, so that a partial vacuum is created and the lungs fill with air, the expansion being followed by relaxation, in which the air is expelled.

BREATHING.

The first portion of this action, known as **inspiration,** is brought about by the contraction of the intercostal and related muscles, by which the ribs are raised, and by the contraction of the dome-shaped diaphragm, by which the posterior wall of the thorax is flattened, and incidentally the abdominal viscera displaced backward. Both actions tend to enlarge the thoracic space. The action of the diaphragm is controlled directly by the phrenic nerves, but all respiratory movements are dependent upon the cervical and thoracic spinal nerves, and there is also a respiratory control centre in the medulla. The expulsion of air, **expiration,** is accomplished without muscular contraction.

FIG. 45. The larger terminal ramifications of the left inferior bronchial ramus, from the dorsal surface; metallic cast of the interior; cf. Figs. 44 and 83, i.

Respiration as a general function is common to all organisms. Though always constructed for easy diffusion, the organs by which the function is discharged differ profoundly in the various groups. This is true even within the limits of the vertebrates, where lower forms are characterized by **gills** for aquatic respiration, and the higher forms by **lungs** for air respiration. The occurrence of a great variety of intermediate and transitional growth stages, in which gills are replaced by lungs, with no modification from one to the other, introduces a peculiar condition into the history of vertebrates. This condition is characterized by the appearance of gill structures in the embryos of all higher forms (Fig. 22) and by the gradual elaboration in the series of air sacs from a simple type, as illustrated in the frog, to the greatly branched lung tubes of mammals.

LUNGS AND GILLS.

THE VASCULAR SYSTEM.

In the rabbit, as in all vertebrates, the **vascular system** (Fig. 46) embraces a central, muscular organ of propulsion, the **heart**, and a series of branched tubes, the **blood-vessels**, the latter being of three different kinds: (a) thick-walled, elastic, distributing vessels — **arteries**; (b) microscopic terminal canals in the peripheral organs—**capillaries**; and (c) thin-walled collecting vessels—**veins.**

PLAN OF THE CIRCULATION.

The chief mammalian feature in this system consists in the division of the heart into two portions, each consisting of a receiving chamber, or **atrium**, and a driving chamber, or **ventricle**, and the arrangement of their vascular connections in such a way that two complete circulations are established. One of these is the **long**, or **systemic circulation.** It is concerned with the distribution of blood to the various parts of the body, with the exception of the lungs. It is established by the left ventricle, the aorta, the carotid and subclavian branches of its arch, and the parietal and visceral branches of its thoracic and abdominal portions. The blood is collected from the anterior portions of the body through paired internal and external jugular and subclavian veins, communicating with the right atrium of the heart through paired superior cavals; from the posterior portions of the body through the unpaired and also asymmetrical inferior caval vein, the latter passing forward on the right of the median plane and entering the posterior end of the right atrium. The second, **short**, or **pulmonary circulation**, is concerned with the distribution of the blood to the lungs for purposes of aeration. It is established by the right ventricle, the pulmonary artery and its paired branches, and by the capillaries of the lungs. The blood is delivered to the left atrium through several pulmonary veins.

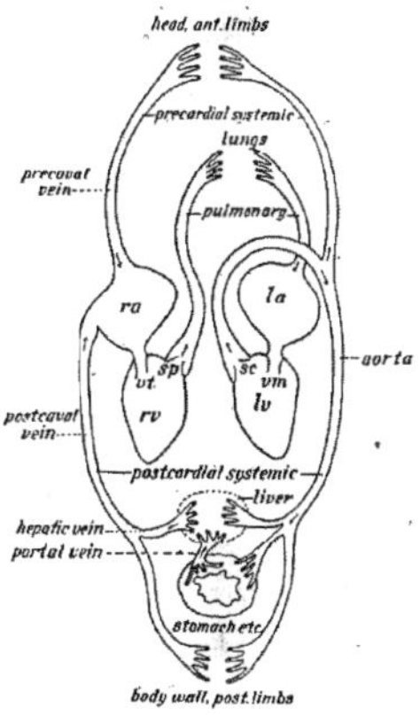

FIG. 46. The mammalian circulation. rv, lv, right and left ventricles; ra, la, right and left atria; sc, sp, semilunar valves of aorta and pulmonary artery; vt, vm, right and left atrioventricular valves.

A similar division of the circulatory organs occurs as a homoplastic modification in birds, which, it will be observed, are also warm-blooded vertebrates.

In general, the blood which is distributed to the various parts of the body passes through but one set of capillary vessels, and is then returned through the systemic veins to the heart. In all vertebrates, however, a special portion of the systemic venous circulation is set aside as the **hepatic portal system** (Fig. 42), distinguished by the possession of a second series of capillary vessels ramifying in the liver. Thus, in the rabbit, the blood distributed to the stomach, spleen, and intestine through the coeliac, superior and inferior mesenteric arteries, is collected into a main intestinal vessel, the **portal vein,** and the latter, approaching the liver through the lesser omentum, divides in that organ into a series of **portal capillaries.** The portal capillaries, like the systemic capillaries proceeding from the hepatic artery, unite in the tributaries of the hepatic veins. In lower vertebrates, although not in the mammalia, a second system of venous capillaries occurs in connection with the kidneys and is known as the **renal portal** system.

The ultimate function of the vascular system is connected with interchange of materials in the tissues. This is brought about through the medium of microscopic **capillaries,** the gross parts of the system being concerned with transportation of materials from one part of the body to another.

FUNCTIONS OF CIRCULATORY SYSTEM.

The propulsive action of the heart is muscular and rhythmic, contraction, or **systolic phases,** alternating with expansion, or **diastolic phases** (cf. p. 63). The flow is maintained in one direction principally by atrioventricular valves of the heart, and by the semilunar valves the of the aorta and pulmonary arteries, though there are also valves in the course of some of the veins. The arteries are tubes with thickened elastic walls. They are expanded by the impulse of blood from the heart, contraction of which is followed by a **pulse wave** in the arteries. The passage of blood into the capillaries takes place more slowly and uniformly, while the arteries contract to their previous diameter. In the return of the blood the veins are largely passive, acting merely as closed channels connecting the

capillaries with the heart. The control of the muscular action of the heart and arteries through the vagus and sympathetic nerves is an important element in maintaining pressure and tone in the vascular system (cf. p. 62).

Many of the peculiar features of the mammalian circulation which at first sight do not appear to be general, but are so in reality, depend on the circumstance that the complete partition of the organs is a final stage of a general progressive development, observable in air-breathing vertebrates, in which the lungs and their vascular connections become perfected for **pulmonary respiration.** On the other hand, the vascular system as it appears in the embryo, more especially its aortic portion, is arranged according to the type of **branchial respiration** as found in fishes. In this condition the blood is sent forward from the heart through a ventral aorta. The latter is connected with a series of paired **branchial aortic arches,** traversing the rudimentary gill structures, and thus passing upward around the sides of the primitive pharynx. The dorsal aorta is formed by the junction of the branchial aortic arches, and passes backward as a main distributing vessel on the ventral side of the axial support. The heart itself is formed primarily on a two-chambered plan, similar to that in fishes, where all the blood is received by single atrium and is delivered forward to the gills by a single ventricle.

PRIMITIVE AORTIC ARCHES.

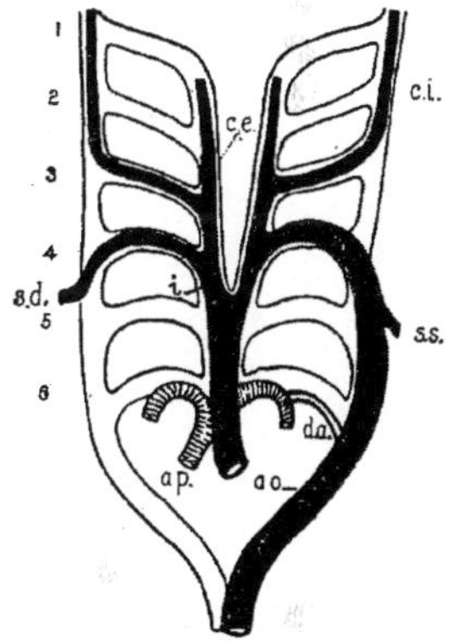

FIG. 47. Plan of the branchial aortic arches. The adult mammalian vessels are indicated in black (systemic) or shaded (pulmonary). 1-6, primary arches; ao., aorta; a.p., pulmonary artery; c.e., external carotid; c.i., internal carotid; d.a., ductus arteriosus (Botalli); i., innominate artery; s.d., right subclavian; s.s., left subclavian. (From Weber, after Boas.)

The definitive condition of the chief arterial vessels is arrived at by an extensive modification of the branchial plan. As indicated in the accompanying diagram (Fig. 47) the arch condition is retained by the aorta and by the pulmonary artery, and it is interesting to note also that the primary connections of these vessels, represented in the embryo by an open canal, the ductus arteriosus (Botalli), is

indicated in the adult condition by a short fibrous cord, the **arterial ligament.** The adult aortic arch represents only the left one of a pair, and since that of the right is only represented imperfectly by the innominate artery and the base of the subclavian of that side, a condition of asymmetry results, which is mainly expressed by the sinistral position of the arch with reference to the oesophagus (Plate VII). By comparison with the embryonic plan, it is seen that the primitive features of the heart and the arterial vessels include the ventral position of the heart itself, the equivalence of the two atria and of the two ventricles—these structures being partitioned internally but imperfectly divided externally—the forward position of the first portion of the aorta, and the position of the aorta as a median vertebral trunk.

The vascular system is noteworthy for several departures from the condition of symmetry, one of these having already been mentioned. In addition, it is seen that in a mammal, as in terrestrial vertebrates generally, the base of the pulmonary artery (Fig. 44) is rotated in a spiral fashion about the base of the aorta, so that from its beginning on the right ventricle it passes across the ventral surface of the base of the aorta, and divides on the dorsal side of the latter into its two main branches. Moreover, the separation of the ventricular portion of the heart into two chambers is associated with an enormous increase in the muscularity of the wall in the left ventricle, or, in other words, in that portion which is concerned with the larger, systemic circulation. The inferior caval vein (Plate VIII), a highly specialized vessel, is asymmetrical, since from its beginning in the pelvic cavity to its termination on the right atrium it lies wholly to the right of the median plane. The azygos vein of the thorax (Plate VII), a vessel uniting the majority of the paired intercostal veins, and interesting as a remnant of the primitive circulation, is also asymmetrical, since the trunk lies to the right of the bodies of the vertebrae, and is connected at its base with the right superior caval vein.

THE LYMPHATIC SYSTEM.

The **lymphatic system**, both in its functional relation and in origin, is an appendage of the venous portion of the vascular system. The system is an important one, of which, unfortunately, little

may be seen by ordinary dissection, the structures which appear in this way in being the lymph glands, or **lymph nodes**, centres of cell formation, occuring in the course of the conducting vessels. These as superficial structures are found either singly, as in the head and neck, or more or less grouped, as in the axillary and inguinal spaces. As deep structures they are conspicuous in the intestinal mesenteries, and in the walls of the digestive tube, occuring in the latter chiefly as continuous masses of **lymph follicles**, as, for example, in the walls of the sacculus rotundus, the vermiform process, or the tonsil; or, again, as **aggregated lymph follicles** (Peyer's patches) at various points in the intestinal wall.

The conducting portion of the system comprises an extensive series of canals, beginning as **lymphatic capillaries** in peripheral organs, and ending as **lymphatic trunks** which empty into the great veins. The lymphatic trunks of the anterior portion of the body are designated from their association with the corresponding veins as **jugular and subclavian.** They enter the venous system on either side at the point of junction of the internal and external jugular veins or of the common jugular and subclavian (Fig. 82).

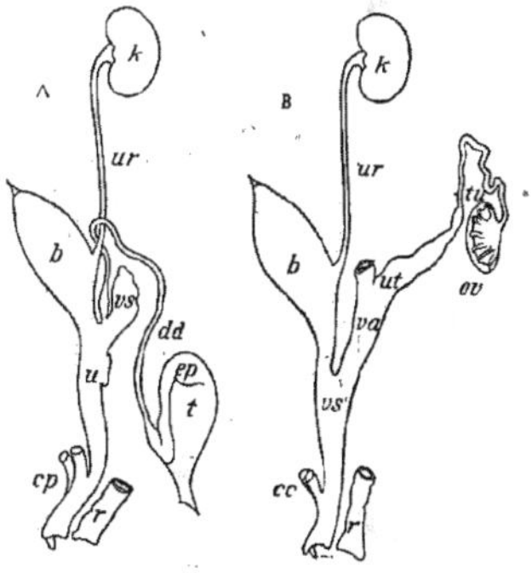

FIG. 48. Homologies of male (A) and female (B) urinogenital systems. b, urinary bladder; cc, crura clitoridis; cp, crura penis; dd, ductus deferens; ep, epididymis; k, kidney; ov, ovary; r, rectum; t, testis; tu, uterine tube; u, urethra; ut, uterus; ur, ureter; va, vagina; vs, seminal vesicle; vs', vestibulum.

The lymphatic vessels of the posterior portion of the body, including the intestine, unite to form a common canal, the **thoracic duct.** The latter lies for the most part between the aorta and the vertebral column, and traverses the thorax in this position to enter the venous system at the same point as the jugular and subclavian trunks of the left side.

The lymphatic capillaries are terminal, absorptive vessels, differing from blood capillaries both in the character of their walls and in their relations to other portions of the system, since they are not interposed as in the vascular system between vessels of a larger order. The lymphatic vessels connecting the capillaries

with the lymphatic trunks form extensive plexuses, in connection with which the lymph nodes are distributed.

THE URINOGENITAL SYSTEM.

The **urinogenital system** comprises two primary systems—**reproductive** and **urinary**—differing widely in their central organs, but associated to a certain extent by having common ducts. In the rabbit, as indicated in the accompanying diagram (Fig. 48), this association extends only to the presence in the two sexes of a urinogenital canal, or **urinogenital sinus** connecting both urinary and genital structures with the outside of the body. This canal is designated in the male as the **urethra**, but in the female as the

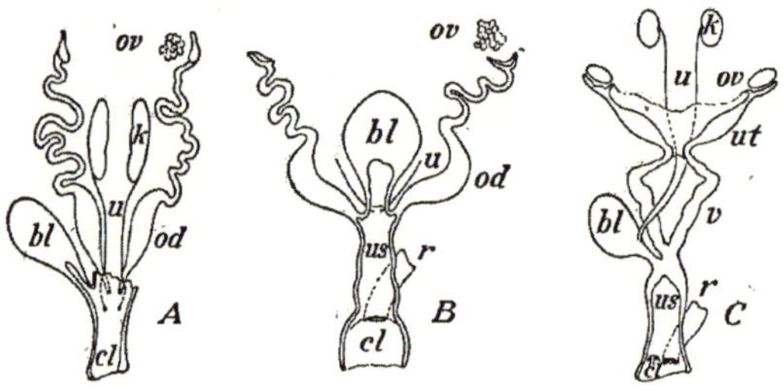

FIG. 49. The principal stages in specialization of the female urinogenital ducts in vertebrates. A, frog; B, monotreme; C, marsupial. bl, bladder; cl, cloaca; k, kidney; od, oviduct; ov, ovary; r, rectum• u, ureter; us, urinogenital sinus (vestibulum); ut, uterine tube; v, vagina. Chiefly from figures of Gegenbaur and Wiedersheim.

vestibulum, since the structure known from the human relation as the female urethra is only a urinary canal leading from the bladder, and in man is not associated with the reproductive ducts.

URINOGENITAL DUCTS IN VERTEBRATES.

In primitive vertebrates (Fig. 49), the urinary and genital ducts open into the posterior end of the digestive tube, the latter forming in this relation a common canal, the **cloaca.** In terrestrial vertebrates, the urinary bladder is developed as a ventral outgrowth of the digestive tube, and, except in amphibians, both sets of ducts undergo a migration from their original position on to the wall of its canal, the latter being thus trasfnormed into a

urinogenital sinus. This development reaches its extreme in the higher mammalia, where the urinogenital sinus is completely separated from the digestive tube, and where the urinary ducts are also transferred from a posterior or **hypocystic** position on the wall of the urinogenital sinus to an anterior or **epicystic** position on the dorsal wall of the bladder.

THE KIDNEYS.

The chief organs of the urinary system are the **kidneys**. They are paired organs, lying against the dorsal abdominal wall, approximately in the position of the embryonic intermediate cell mass from which they are formed. That of the left side is displaced backward, out of the position of symmetry, on account of the posterior development of the greater curvature of the stomach. The kidneys appear as solid organs, brownish in colour and bean-like in general shape, enclosed by a fibrous coat, and connected medially with the expanded end of the ureter. In the rabbit the kidney appears as an almost continuous mass, in which, however, slight traces of **lobulation** can be distinguished. In many mammals, such as sheep and bear, the organ is composed of distinct and separable lobules. This condition is clearly shown in the human kidney in foetal life, and though much more concentrated in the adult, the lobulated condition appears internally in the division of the ureter into several **renal calyces,** each of them connected with a corresponding renal papilla.

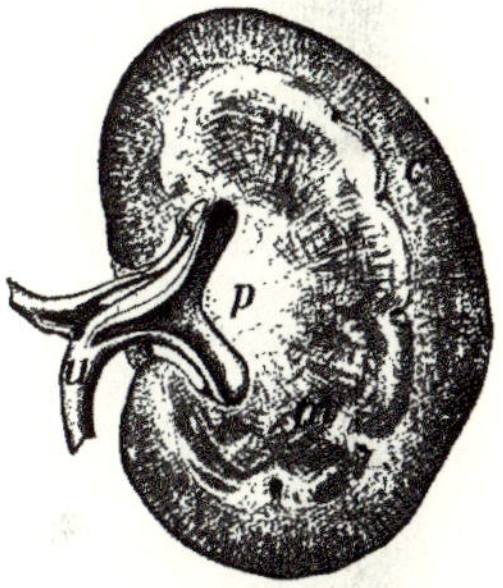

FIG. 50. The left kidney, divided horizontally lengthwise, cut surface of dorsal half. c, cortical substance; m, medullary substance; p, renal papilla; u, ureter.

FORM IN MAMMALS.

When horizontally divided (Fig. 50), the kidney is seen to be made up of a more vascular and granular external layer, termed the **cortex,** and of a somewhat radially striated, central mass, termed the **medulla.** In the rabbit, there is a single **renal papilla**, and the expended end or **pelvis** of the ureter is undivided. Notwithstanding the solid appearance of cortex and medulla, the kidney is made up of a system of **tubules**, the relation of which to the vascular system and to the outside of the body is such that fluid materials

to be excreted may pass into them from the blood stream. The primary tubule (Fig. 51A) begins in each case in the cortical substance with a cup-like structure, known as a **renal** or Malpighian **corpuscle,** containing a network or **glomerulus** of minute vessels from the branches of the renal artery. The blood stream is thus separated only by a thin membrane from the cavity of the tubule, the wall of which is complete. The terminal parts of the tube system have a characteristic course in the kidney substance, which accounts for the difference in appearance as between the cortex and medulla, and are ultimately connected with common **collecting tubules** opening on the surface of the papilla. The excreted fluid, urine, contains characteristic nitrogenous waste materials, usually urea, but in some cases hippuric acid. These materials are formed in the liver and perhaps elsewhere in the body.

INTERNAL STRUCTURE AND FUNCTION.

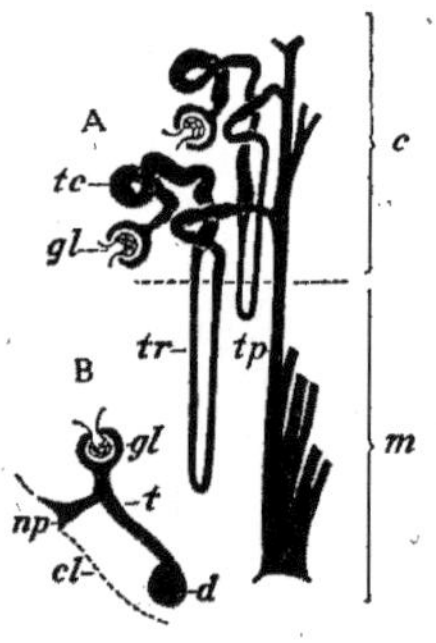

FIG. 51. Kidney tubules. A, plan of arrangement in adult mammal: c, cortex; m, medulla; gl, vascular glomerulus; tc, proximal convoluted portion of tubule; tr, proximal straight portion of tubule (Henle's loop); tp, collecting tubule to renal papilla. From Weber, after v. Ebner. B, plan of a single primitive kidney tubule in lower vertebrates, glomerulus; np, nephrostome; cl, coelomic epithelium; t, body of tubule; d, longitudinal duct.

Like all other parts of the urinogenital system, the structure and embryonic development of the mammalian kidney affords a remarkable illustration of the extent to which the adult form of an organ is dependent upon ancestry. In the vertebrate phylum, three pairs of kidneys have been recognized. They occur in antero-posterior order in the body, either in embryo or adult, they are of increasing specialization, and their order of appearance and functional value are directly associated with the degree of general specialization of the groups in which they occur. These organs have been designated as **pronephros, mesonephros,** and **metanephros.** The **metanephric** kidney is characteristic of mammals, while the mesonephros is embryonic. The latter is, however, the adult kidney in the frog and allied animals, its duct in the male serving both as reproductive duct and ureter. The

HOMOLOGIES OF VERTEBRATE KIDNEYS.

presence of this kidney in embryo mammals, together with its duct, determines the form and connections of the ductus deferens with the terminal portions of the urinary system. The pronephros, on the other hand, is a vestigial kidney in all vertebrates, but its duct

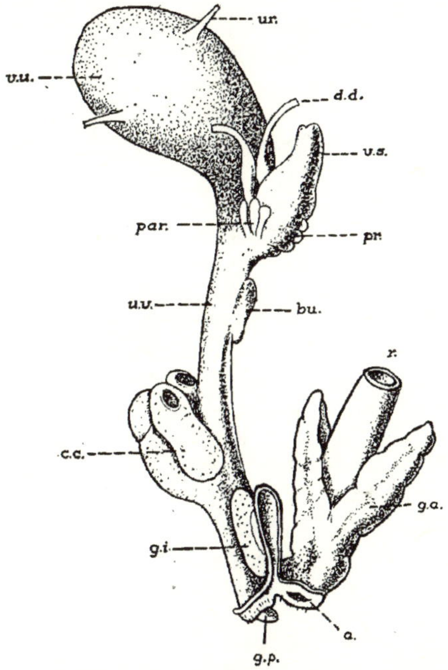

FIG. 52. The male urinogenital ducts and related structures, viewed from the lateral surface. After Rauther: a., anal aperture; bu., bulbourethral gland; c.c., corpus cavernosum; d.d., ductus deferens; g.a., anal (rectal) gland; g.i., inguinal gland; g.p., glans penis; par., paraprostatic glands; pr., prostate; r., rectum; ur., ureter; u.v., urethra (membranous portion); v.s., seminal vesicle; v.u., urinary bladder.

system, open proximally to the body cavity plays an important part in the formation of the oviduct in the female of all classes of vertebrates. The development of the kidney reveals a condition, also suggested in a less perfect way by the mesonephros, in which it is shown that the primary connections of the kidney tubules are

with the coelomic cavity (Fig. 51 B). Their ciliated internal openings, termed **nephrostomes,** are not developed in specialized kidneys, but may be seen, even in the adult condition, in some lampreys, where they communicate with that portion of the body cavity enclosing the heart.

THE TESTIS AND ITS DUCT.

In the rabbit, as in all mammals, the male gonad or **testis** (Fig. 55), in which the male germ cells, **spermatozoa,** are matured, is connected with the peripheral duct system (Fig. 52) by means of the **epididymis** and the **ductus deferens,** parts of the mesonephric connections of the embryo. While the ductus deferens is a single tube, the epididymis consists of an aggregation of small tubules, lying chiefly toward the anterior end of the testis, but with the tubules not individually discernible. The testis is formed in the embryo in all vertebrates in association with the dorsal abdominal wall, but in many mammals moves backward to a scrotal position, either periodically or permanently in the course of development. This change, known as the **descent** of the testis, is brought about through the agency of a muscular cord, the **gubernaculum,** the connections of which in the rabbit, are retained in the adult condition. The migration of the organ determines a number of peculiarities in the relations of its bloodvessels and peritoneal connections.

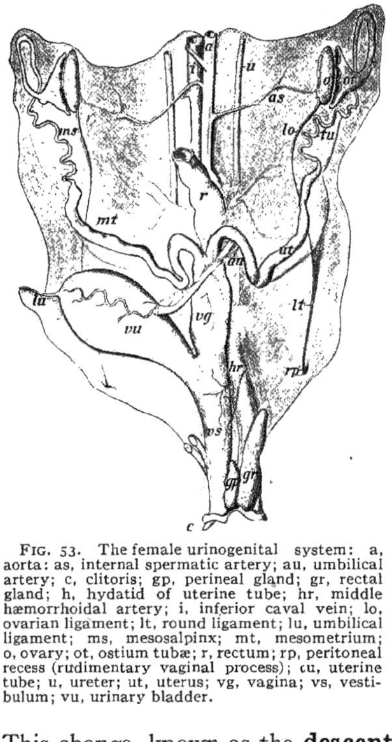

FIG. 53. The female urinogenital system: a, aorta: as, internal spermatic artery; au, umbilical artery; c, clitoris; gp, perineal gland; gr, rectal gland; h, hydatid of uterine tube; hr, middle hæmorrhoidal artery; i, inferior caval vein; lo, ovarian ligament; lt, round ligament; lu, umbilical ligament; ms, mesosalpinx; mt, mesometrium; o, ovary; ot, ostium tubæ; r, rectum; rp, peritoneal recess (rudimentary vaginal process); tu, uterine tube; u, ureter; ut, uterus; vg, vagina; vs, vestibulum; vu, urinary bladder.

The female gonad or **ovary** (Fig. 53) lies on the dorsal wall of the abdominal cavity, thus retaining to a large extent the primitive

position. Its gubernacular connections are, however, plainly discernible in the adult animal by the **ovarian** and **round ligaments,** the latter being inserted into a small pocket of the abdominal wall simulating the testis sac.

THE OVARY AND OVIDUCTS.

Though inconspicuous in gross size as compared with the testis, the ovary is concerned with the formation of cells of relatively large dimensions, the female germ cells or **ova.** These undergo their principal development as single cells in the tissue of the organ, but at times, through rupture of the enclosing **follicles,** they gain access to the surface, and thence pass directly into the open mouth of the **uterine tube.** If fertilized, they begin their segmentation and further development into an embryo, the latter becoming attached to the wall of the **uterus.** A **placental connection** is formed by which nourishment is carried to the embryo, during the period of intra-uterine life, in the rabbit about thirty days. There are two complete uteri, the cavities of which are connected distally with the unpaired vagina, and through this with the urinogenital sinus. The size and appearance of the uteri depend upon the age of the animals examined, and upon whether or not they are pregnant or have borne young. The uteri of pregnant females are greatly enlarged and vascular. They contain from five to eight young, the position of which may be easily seen from the expansion of those parts of the tubes in which they lie.

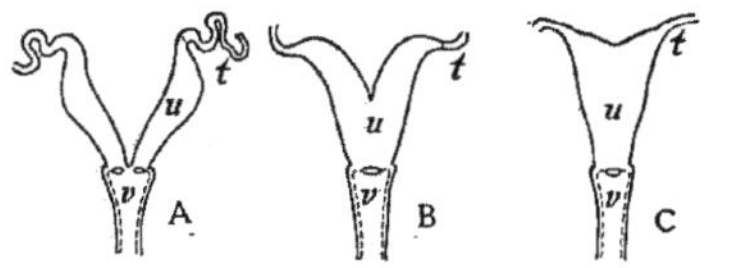

FIG. 54. Three stages of specialization of the uterus. A, uterus duplex; B, uterus bicornis; C, uterus simplex. t, uterine tube; u, uterus; v, vagina.

The paired condition of the uteri in the rabbit is especially instructive because of its primitive nature as compared with that in many other mammals. **Paired oviducts** (Fig. 49A) are the rule in lower vertebrates, where the function is simply to carry the eggs to the outside of the body, This condition is retained with minor modifications to the mototreme stage of mammals, but in higher forms of the latter the ducts are successively coalesced. In marsupials there are still paired vaginae, while in placentals the structure is unpaired. In the rabbit, as in many lower placentals,

there are two complete uteri, and as an organ the whole structure represents the stage of **uterus duplex** (Fig. 54 A). A partly fused condition existing in some mammals, for example sheep, is known as **uterus bicornis** (Fig. 54 B), while the completely fused condition in man is known as **uterus simplex** (Fig. 54 C). It is characterized by the independent opening of the two uterine tubes into a single uterine cavity. The successive stages of coalescence are doubtless associated with progressive reduction of the number of young, the success of the species being determined by greater perfection of the placental apparatus.

THE SEROUS CAVITIES.

The organs collectively described as **visceral** are those associated with the serous cavities They belong to several systems, but present the common feature of being projected into the membranous linings of these cavities so that they are more or less completely invested by them.

The **serous sacs** are extensive body-spaces, derivatives of a primary body cavity or **coelom.** They are usually considered as containing the visceral organs, but the condition is more accurately described as one in which the visceral organs encroach, chiefly from a dorsal position, on the enclosing membranes. The latter are thus divided into two portions, one of which is distributed as a **parietal** or peripheral layer, forming the enclosure of the sac, while the other is disposed as a visceral layer on the surface of the visceral organs. The serous sacs are enclosed by thin, moist, **serous membranes,** consisting chiefly of mesothelium, which give to the visceral organs their characteristic appearance.

In lower vertebrates, where the diaphragm is absent or imperfectly developed, the coelom is divided into two chief portions—the **pericardial cavity,** enclosing the heart, and the **pleuroperitoneal cavity,** lodging the remaining visceral organs, including in terrestrial vertebrates the lungs. In the mammalia the pleuroperitoneal cavity is completely divided into two portions by the diaphragm, the smaller pleural portion being again divided into **right** and **left pleural cavities** through the presence of certain structures filling the median portion of the thorax. There are thus recognizable in a

mammal four large serous spaces, namely, the pericardial peritoneal, and paired pleural cavities.

The **pericardial cavity,** the smallest of these spaces, is situated between the paired pleural cavities. Its enclosing membrane, the **pericardium,** forms a capacious sac for the heart, and is reflected directly over the surface of the latter as a thin membrane, the **epicardium.**

The **pleural cavities** are those lodging the lungs, the latter being projected into them from a medial position. The lining membrane or **pleura** is divided into three chief portions—the **pulmonary pleura,** investing the greater part of each organ, the **costal pleura,** lining the internal surface of the thorax, and the **diaphragmatic pleura,** covering the anterior surface of the diaphragm. The latter is broadly connected with the pulmonary pleura through the **pulmonary ligament.**

The **peritoneal cavity,** the largest of the serous spaces, comprises in a mammal a general portion, the **abdominal cavity,** and its posterior extension into the pelvis—in the male also into the **sac of the testis.** The general relation of the cavity to the abdominal organs is indicated diagrammatically in Fig. 21. Its lining membrane, the peritoneum, is divisible into two portions, the **parietal peritoneum,** lining the abdominal wall, and the **visceral peritoneum,** investing the visceral organs. Of the latter the kidneys encroach only to a minor extent on the serous lining, so that they are covered by peritoneum only on their ventral surfaces. The digestive tube, on the other hand, is removed to such an extent from the abdominal wall that the peritoneum forms a complete serous coat, and is connected with the parietal peritoneum of the wall through a thin transparent membrane, the **mesentery.** The latter consists of two plates of peritoneum, enclosing between them a thin layer of connective tissue, the **lamina mesenterii propria,** for the transmission of nerves, bloodvessels and lymph canals.

As indicated above, the relations of the abdominal portion of the digestive tube are greatly modified by its elongation and displacement from a median position. Thus, while in the embryo the **common mesentery** (Fig. 40) is recognizable as a continuous median vertical fold, in the adult it follows the convolutions of the

digestive tube, and is therefore considered as divided into corresponding parts. In many cases the relations of these are greatly complicated by secondary adhesions. In the rabbit the **mesoduodenum, mesentery,** and **descending mesocolon** will be recognized as parts in which a more typical arrangement is retained. Moreover, in the anterior portion of the abdominal cavity the peritoneum is concerned not only with the investment of two large visceral structures, the stomach and the liver, but also with the formation of a lining for the posterior surface of the diaphragm. Thus the general condition is less simple than in the small and large intestines. The peritoneum, passing from the dorsal wall, successively invests the spleen, the stomach, and the liver, and is reflected from the last-named structure to the diaphragm and the ventral body-wall through the **coronary, triangular,** and **falciform ligaments.** Its gastric portion is differentiated into the **mesogastrium** (phrenicosplenic and gastrolienal ligaments), the **greater omentum,** and the **lesser omentum.** Similarly, in the posterior part of the body the peritoneum passes from the rectum to the urinary bladder, enclosing also in the female the vagina. It is then reflected to the ventral body-wall as the **middle umbilical fold.**

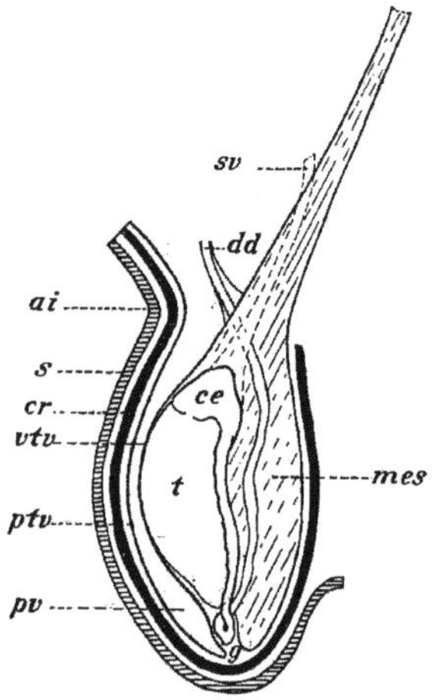

FIG. 55. Diagram showing the relation of the testis to its investments: a.i., inguinal ring; c.e., caput epididymidis; cr., cremaster muscle; dd, ductus deferens; g, gubernaculum; mes, mesorchium; ptv, and vtv, parietal and visceral layers of the tunica vaginalis propria; pv, cavity of the vaginal process; s, integument of the scrotum; sv, spermatic vessels; t, testis.

In the male, as indicated in the accompanying diagram (Fig. 55), the peritoneal relations of the testis are greatly modified by the migration of the organ from an abdominal to a scrotal position. The entire sac lodging the testis is an evaginated portion of the abdominal wall, and since in the rabbit the cavity is widely open throughout life to the abdominal cavity the lining membrane—that designated as the **parietal layer** of the **tunica vaginalis propria**—

is continuous with the parietal peritoneum of the abdomen, and thus represents a permanent **vaginal process.** Like other structures of the abdominal cavity, the testis itself is covered by peritoneum, the latter being designated as the **visceral layer** of the **tunica vaginalis propria.** This investment is connected with the parietal layer by the **mesorchium,** and in the rabbit it will be observed that the latter is chiefly attached forwards on the dorsal wall of the abdomen, *i.e.*, in a position indicating the original situation of the testis itself.

In the female the ovary is closely associated with the dorsal wall of the abdomen, and its supporting peritoneum, the **mesovarium,** is insignificant. Its duct in passing backward, however, becomes greatly displaced from a dorsal position, and thus comes to be supported by a broad fold of peritoneum. The latter is considered to consist of two portions, one, the **mesosalpinx,** being the support of the uterine tube, the other, the **mesometrium,** that of the uterus. The entire fold, however, forms a continuous structure, and is known in this relation as the **broad ligament.**

REGIONAL SECTIONS.

The following plate-figures (I-VIII) are from characteristic sections of a rabbit-foetus of 56mm., and may be used either in connection with the general features of topography as outlined above, or for the identification of various minor structures appearing in the dissection.

Certain points regarding the sections are perhaps worthy of notice. First, in the longitudinal section illustrated in Plates I and II it will be noticed that paired structures frequently appear; this being because of the fact that the section is not exactly median, at least in certain places. Second, in using sections of the foetus for gross anatomical features it is necessary to make allowance in some cases for the different proportions of organs, and consequent slight differences in position, in the foetal as compared with the adult condition. Finally, many of the features appearing in the original sections are such as could not be reproduced in the plates, although they are indicated in the accompanying skeleton figures, and may be referred to in this way.

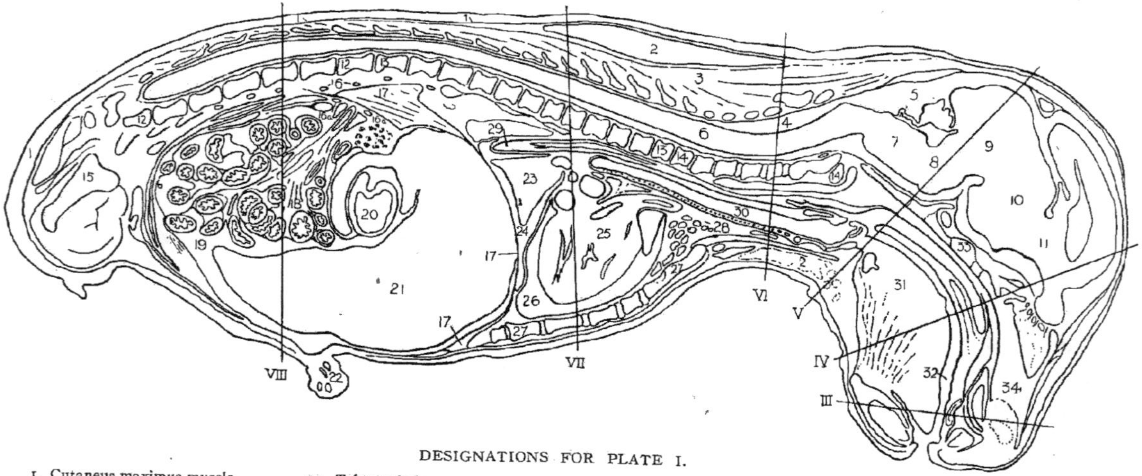

DESIGNATIONS FOR PLATE I.

1. Cutaneus maximus muscle.
2. Fat-masses of neck.
3. Spinal musculature.
4. Vertebral canal.
5. Cranial cavity.
6. Spinal cord.
7. Myelencephalon.
8. Metencephalon.
9. Mesencephalon.
10. Diencephalon.
11. Telencephalon.
12. Lumbar vertebrae.
13. Thoracic vertebrae.
14. Cervical vertebrae.
15. Pelvis.
16. Ganglia of sympathetic trunk.
16A. Superior mesenteric ganglion (prevertebral).
17. Diaphragm.
18. Intestines.
19. Abdominal cavity.
20. Stomach (pyloric portion).
21. Liver.
22. Umbilicus (base of cord and vessels).
23. Lung.
24. Pleural cavity.
25. Heart and aorta.
26. Pericardial cavity.
27. Sternum.
28. Thymus gland.
29. Oesophagus.
30. Trachea.
31. Tongue.
32. Oral cavity.
33. Basicranium.
34. Nasal septum.

III-VIII indicate levels of corresponding transverse sections.

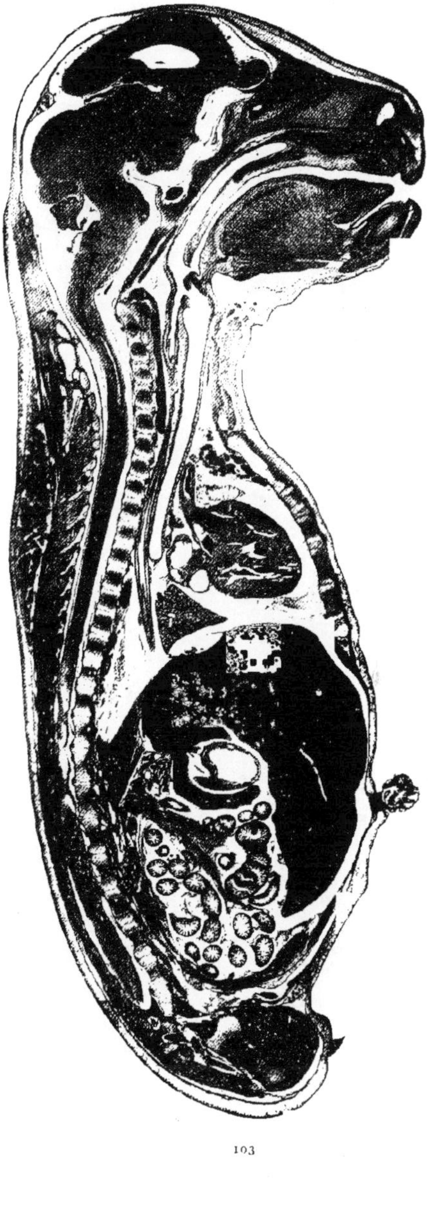

I.

A MEDIAN VERTICAL SECTION OF THE ENTIRE BODY.

DESIGNATIONS FOR PLATE II.

1. Transverse sinus of dura mater.
2. Dura mater.
3. Pallium of cerebral hemisphere.
4. Lateral ventricle.
5. Olfactory bulb.
6. Olfactory tract.
6a. Divided olfactory nerve in the cribriform plate.
7. Chorioid plexus of third ventricle.
8. Anterior commissure.
21. Posterior medullary velum.
22. Cervical flexure.
23. Central canal of spinal cord.
24. Hypophysis.
25. Frontal bone.
26. Nasal bone.
27. Nasal fossa.
28. Mesethmoid cartilage.
29. Cartilage of vomeronasal organ.
30. Premaxilla.
31. Nasopalatine duct and cartilage.
41. Oral portion of pharynx.
42. Epiglottis and epiglottic cartilage.
43. Thyreoic cartilage of larynx.
44. Laryngeal cavity.
45, 45a. Cricoid cartilage.
46. Oesophagus.
47. Cricothyreoideus muscle.
48. Thyreoid gland.
49. Sternohyoideus muscle.
50. Genioglossus muscle.
51. Geniohyoideus muscle.

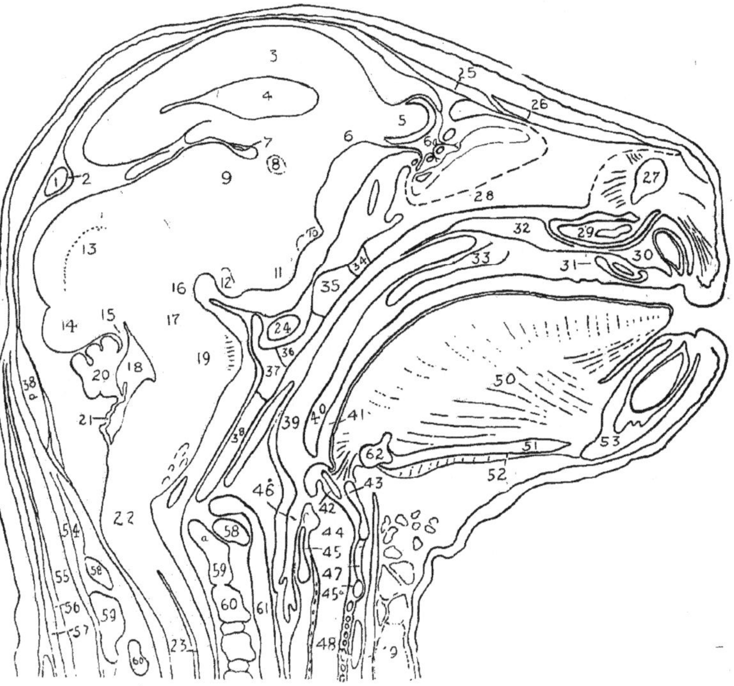

9. Thalamus.
10. Optic chiasma.
11. Tuber cinereum.
12. Mamillary body.
13. Superior colliculus.
14. Inferior colliculus.
15. Anterior medullary velum.
16. Cerebral peduncle; cephalic flexure.
17. Isthmus rhombencephali.
18. Fourth ventricle.
19. Pons; pontine flexure.
20. Cerebellum.
32. Maxilla.
33. Hard palate (palatine and maxilla).
34. Presphenoid.
35. Intersphenoidal synchondrosis.
36. Basisphenoid; hypophyseal fossa.
37. Sphenooccipital synchondrosis.
38. Basioccipital.
38a. Supraoccipital.
39. Nasal portion of pharynx.
40. Soft palate.
52. Mylohyoideus muscle.
53. Mandible.
54. Occipital musculature.
55. Semispinalis capitis.
56. Rhomboideus minor.
57. Superior portion of trapezius.
58. Atlas.
59. Epistropheus.
59a. Odontoid process.
60. Third cervical vertebra.
61. Median vertebral vein.
62. Body of hyoid bone.

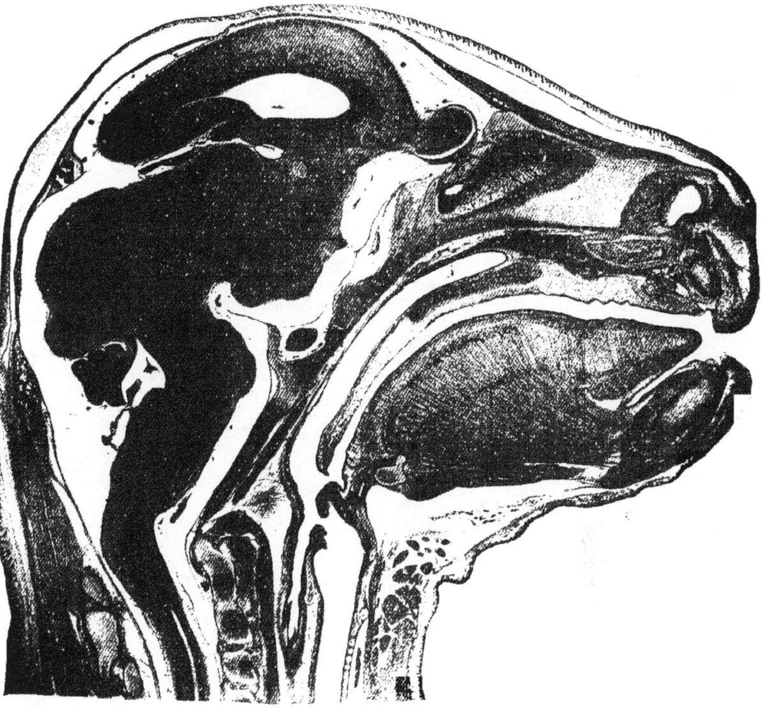

II.

A MEDIAN VERTICAL SECTION OF THE HEAD.

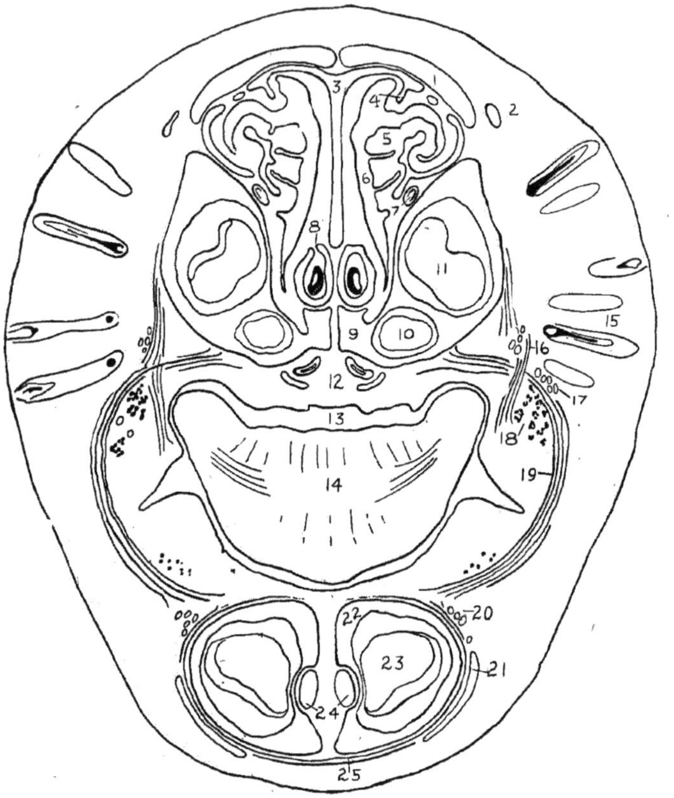

DESIGNATIONS FOR PLATE III.

1. Nasal bone.
2. Levator alae nasi muscle.
3. Nasal septum.
4. Nasoturbinal cartilage.
5. Maxilloturbinal (concha inferior).
6. Nasal fossa.
7. Nasolacrimal duct.
8. Vomeronasal organ and cartilage.
9. Premaxilla.
10. Small upper incisor.
11. Large upper incisor.
12. Nasopalatine ducts.
13. Oral cavity.
14. Tongue.
15. Vibrissae.
16. Caninus muscle.
17. Terminals of superior maxillary nerve.
18. Buccal glands.
19. Buccinator muscle.
20. Terminals of inferior alveolar nerve.
21. Quadratus labii inferioris muscle.
22. Mandible.
23. Lower incisor.
24. Meckel's cartilage (primary mandibular arch).
25. Mentalis muscle.

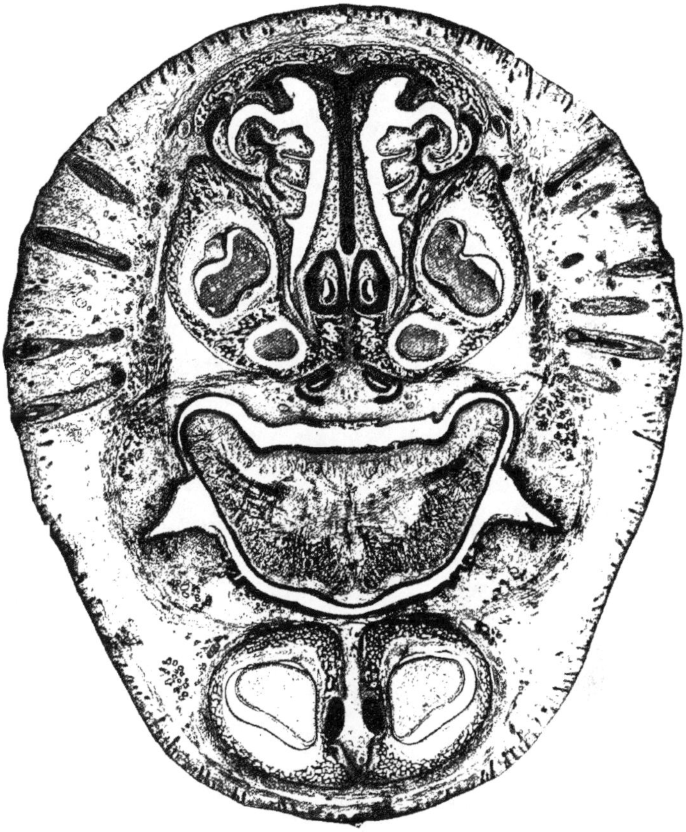

III.
A TRANSVERSE SECTION OF THE ANTERIOR NASAL REGION.

DESIGNATIONS FOR PLATE IV.

1. Superior sagittal sinus of dura mater.
2. Lateral ventricle.
3. Cerebral hemisphere.
4. Pia mater.
5. Frontal bone.
6. Cartilage of orbital wing.
7. Mesethmoid cartilage.
8. Cupula posterior cartilage.
9. Obliquus superior muscle.
10. Ophthalmic vessels and nerves.
11. Levator palpebrae superioris muscle.
12. Rectus medialis muscle.
25. Nasal tract; choana.
26. Palatine bone.
27. Oral cavity.
28. Palatine nerve.
29. Sphenopalatine ganglion.
30. Infraorbital vein.
31. Internal maxillary artery.
32. Maxillary nerve.
33. Maxilla.
34. Zygomatic bone.
35. Submaxillary duct.
36. Buccinator muscle.

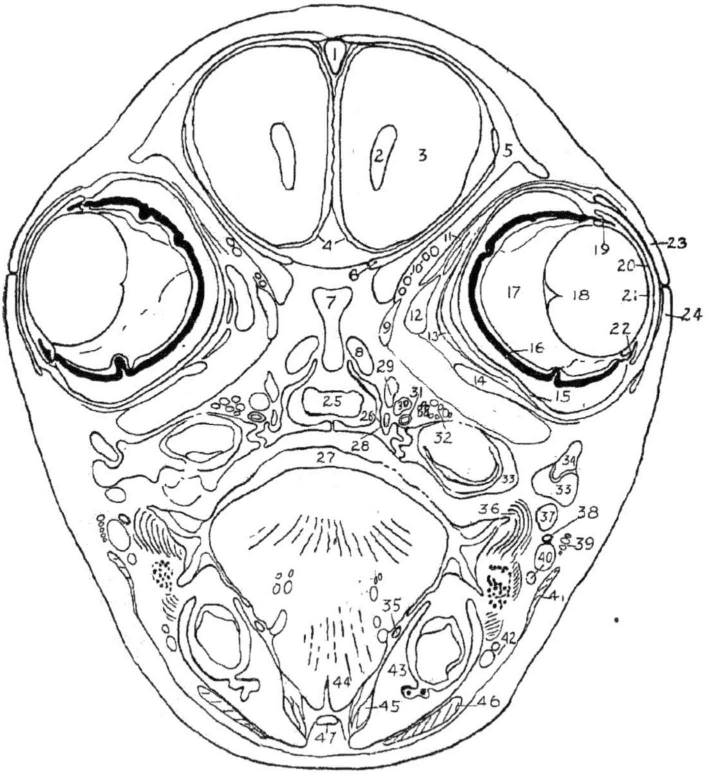

13. Retractor oculi muscle.
14. Rectus inferior muscle.
15. Sclera.
16. Retina and chorioidea.
17. Vitreous body.
18. Lens.
19. Posterior chamber of eye.
20. Anterior chamber.
21. Cornea.
22. Ciliary body and iris.
23. Upper eyelid.
24. Lower eyelid.
37. Masseter muscle.
38. Parotid duct.
39. Facial nerve.
40. External maxillary artery and vein (anterior facial vein).
41. Platysma muscle.
42. Inferior labial artery and vein.
43. Mandible.
44. Genioglossus muscle.
45. Digastricus muscle.
46. Quadratus labii inferioris muscle.
47. Geniohyoideus muscle.

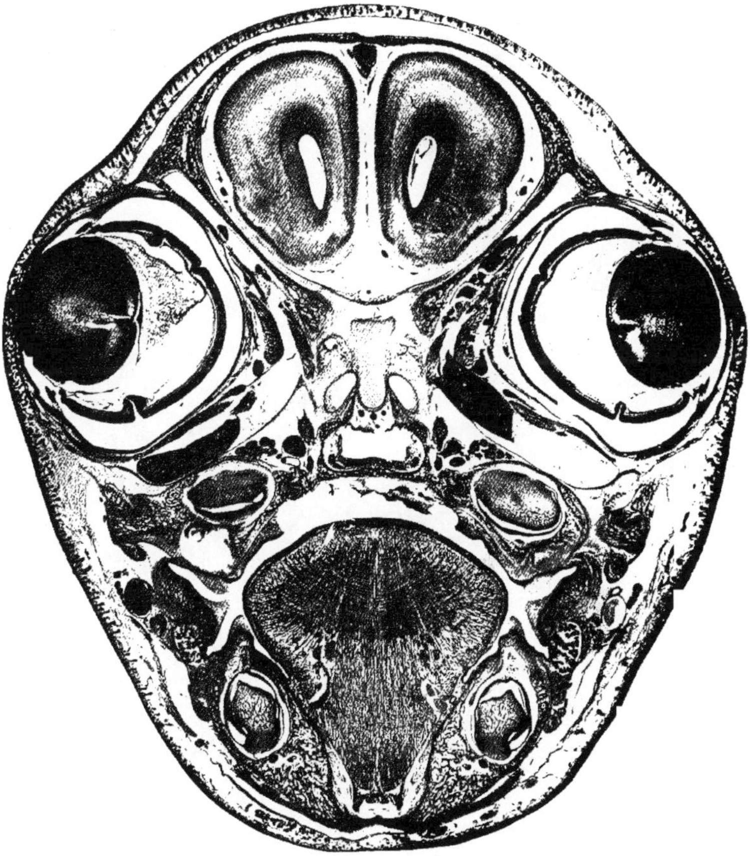

IV.

A TRANSVERSE SECTION OF THE ORBITAL REGION.

DESIGNATIONS FOR PLATE V.

1. Parietal bone.
2. Transverse sinus of dura mater.
3. Superior colliculus.
4. Cerebral aqueduct.
5. Isthmus rhombencephali.
6. Pons.
7. Trigeminal nerve.
8. Basilar artery.
9. Facial nerve.
10. Cartilaginous auditory capsule.
11. Cochlea.
21. Longus capitis.
22. Rectus capitis anterior.
23. Oral portion of pharynx.
24. Thyreohyoideus muscle.
25. Sternohyoideus muscle.
26. Greater cornu of hyoid.
27. Stylohyoideus major muscle.
28. Lingual artery.
29. Hypoglossal nerve.
30. Tendon of digastricus muscle.
31. External maxillary artery.

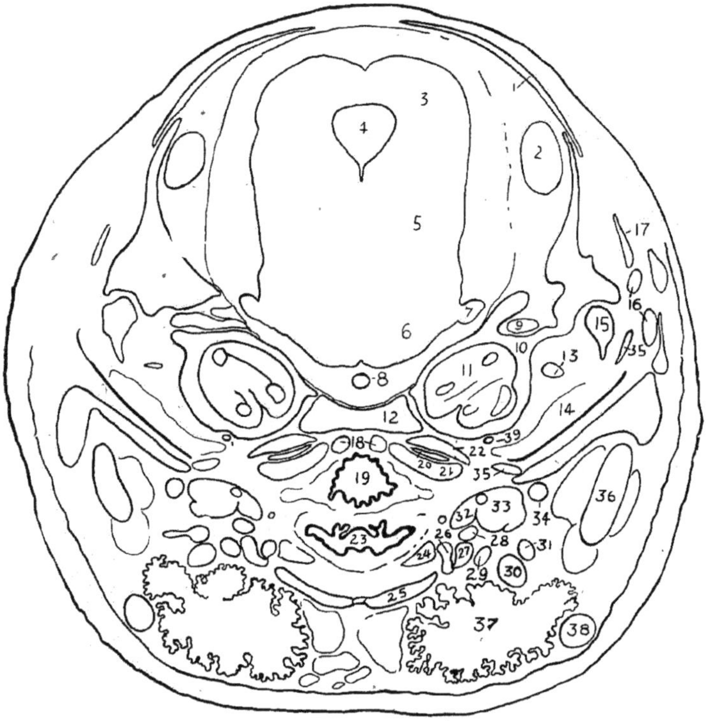

12. Basioccipital bone.
13. Tensor tympani muscle.
14. Tympanic cavity.
15. Malleus.
16. Tributaries of posterior facial vein.
17. Squamosal bone.
18. Cephalic portion of median vertebral vein.
19. Nasal portion of pharynx.
20. Origin of basioclavicularis and levator scapulae major muscles.
32. Stylohyoideus minor.
33. Styloglossus.
34. Internal maxillary artery.
35. Tympanic bone.
36. Mandible.
37. Submaxillary gland.
38. Anterior facial vein.
39. Internal carotid artery.

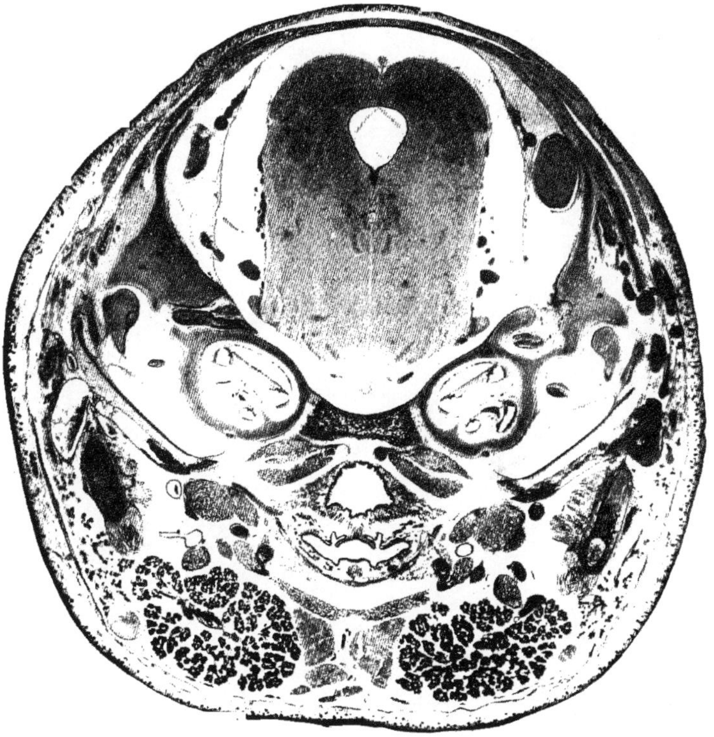

V.

A TRANSVERSE SECTION OF THE AUDITORY REGION.

DESIGNATIONS FOR PLATE VI.

1. Rhomboideus minor.
2. Superior portion of trapezius.
2a. Levator scapulae minor.
3. Splenius.
4. Semispinalis capitis.
5. Rectus capitis posterior superficialis.
6. Obliquus capitis major.
7. Arch of epistropheus.
8. Ganglion of posterior root.
9. Longissimus cervicis.
19. Oesophagus.
20. Inferior thyreoid nerve.
21. Inferior thyreoid vein.
22. Trachea.
23. Thyreoid gland.
24. Cardiac branch of vagus (n. depressor).
25. Sympathetic trunk.
26. Vagus nerve.
27. Common carotid artery.
28. Internal jugular vein.

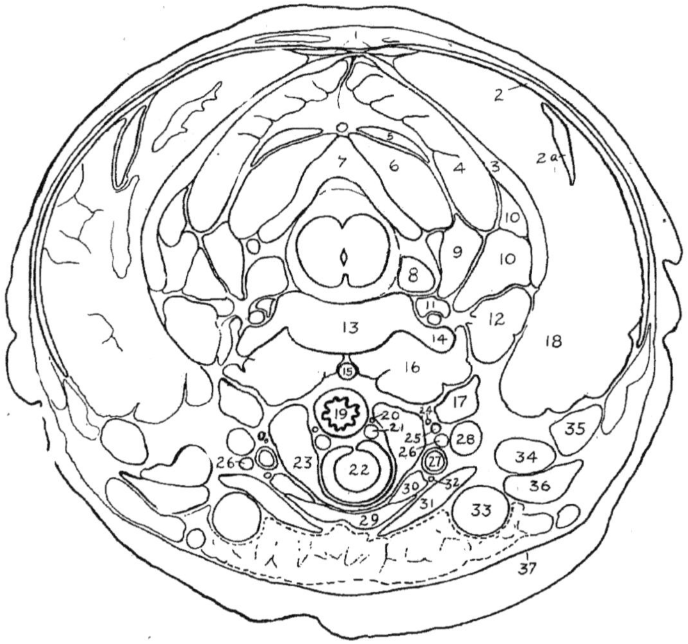

10. Longissimus capitis.
11. Vertebral artery and vein.
12. Longus atlantis.
13. Vertebral body.
14. Tansverse process (anterior root).
15. Median vertebral vein.
16. Longus colli.
17. Longus capitis.
18. Fat-body.
29. Sternohyoideus muscle.
30. Sternothyreoideus muscle.
31. Sternomastoideus muscle.
32. Descending ramus of hypoglossal nerve.
33. External jugular vein.
34. Basioclavicularis muscle.
35. Levator scapulae major muscle.
36. Cleidomastoideus.
37. Platysma.

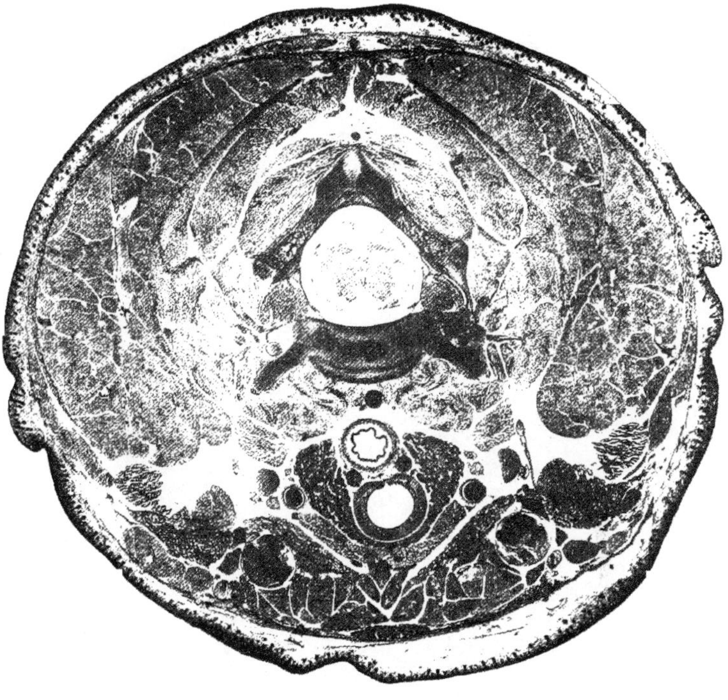

VI.

A TRANSVERSE SECTION OF THE ANTERIOR CERVICAL REGION.

DESIGNATIONS FOR PLATE VII.

1. Semispinalis dorsi.
2. Longissimus dorsi.
3. Iliocostalis.
4. Spinal cord.
5. Ganglion of posterior root and intercostal nerve.
6. Tubercle of rib.
7. Head of rib.
8. Sympathetic trunks.
9. Azygos vein.
10. Thoracic aorta.
11. Oesophagus.
24. Costal pleura.
25. Bone ribs.
26. Costal cartilage.
27. Sternum.
28. Cutaneus maximus muscle.
29. Inferior portion of trapezius.
30. Rhomboideus major.
31. Inferior angle of scapula.
32. Latissimus dorsi.
33. Serratus posterior.
34. Intercostales externi and interni.
34a. Intercostalis internus.

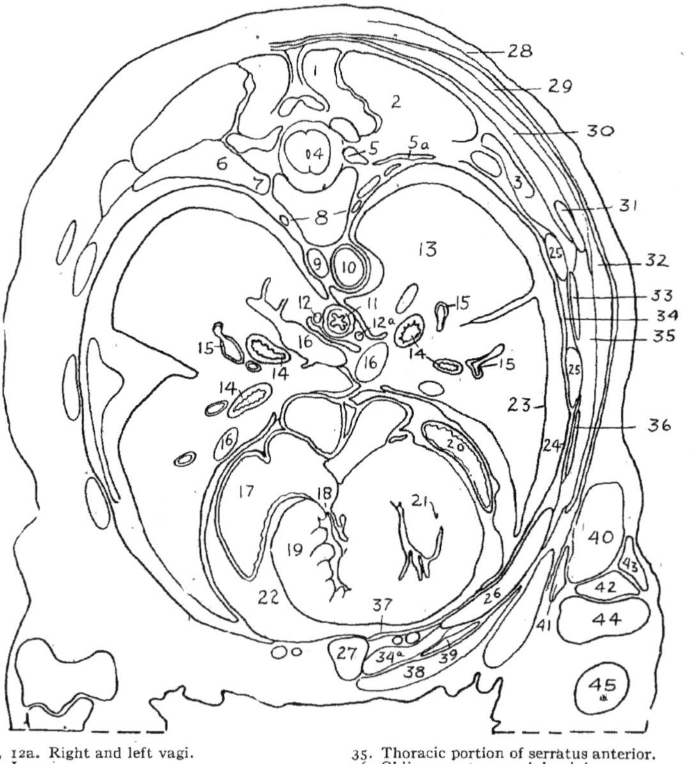

12, 12a. Right and left vagi.
13. Lung.
14. Bronchi.
15. Branches of pulmonary artery.
16. Pulmonary veins.
17. Right atrium.
18. Tricuspid valve.
19. Right ventricle.
20. Left atrium.
21. Left ventricle.
22. Pericardial cavity.
23. Pulmonary pleura.
35. Thoracic portion of serratus anterior.
36. Obliquus externus abdominis.
37. Transversus thoracis.
38. Pectoralis major.
39. Rectus abdominis.
40. Anconaeus longus (caput longum of triceps).
41. Extensor antibrachii parvus.
42. Anconaeus medialis.
43. Anconaeus lateralis.
44. Distal extremity of humerus.
45. Proximal portion of radius.

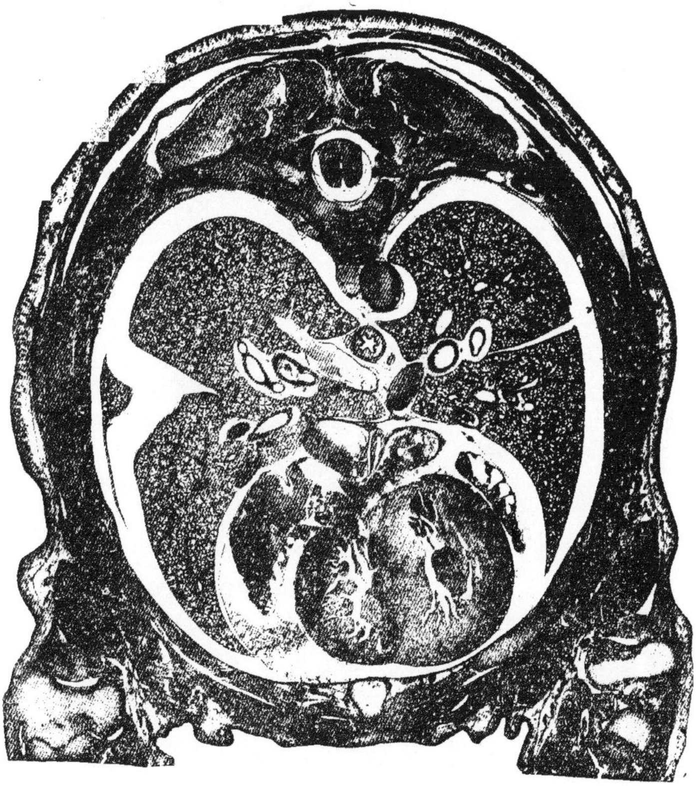

VII.
A TRANSVERSE SECTION OF THE THORAX.

DESIGNATIONS FOR PLATE VIII.

1. Spinal cord.
2. Vertebral canal.
3. Vertebral body.
4. Sacrospinalis muscle.
5. Quadratus lumborum.
6. Psoas major.
7. Psoas minor.
8. Sympathetic trunk.
9. Abdominal aorta.

18, 18a. Posterior and anterior lobules of left lobe of liver.
19, 19a. Right lobe of liver.
20. Obliquus internus abdominis and transversus abdominis.
21. Obliquus externus abdominis
22. Rectus abdominis.
22a. Cutaneus maximus.

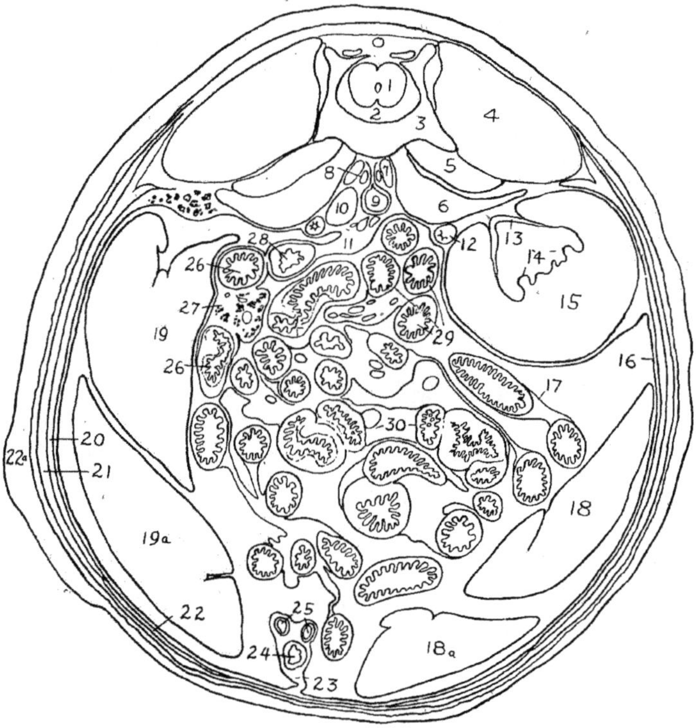

10. Inferior caval vein.
11. Descending mesocolon.
12. Ureter.
13. Renal pelvis.
14. Renal papilla.
15. Left kidney.
16. Parietal peritoneum.
17. Visceral peritoneum.

23. Middle umbilical fold.
24. Urinary bladder (canal of foetal allantois).
25. Umbilical arteries.
26. Duodenum.
27. Pancreas and mesoduodenum.
28. Descending colon.
29. Parts of mesenterial small intestine.
30. Caecum.

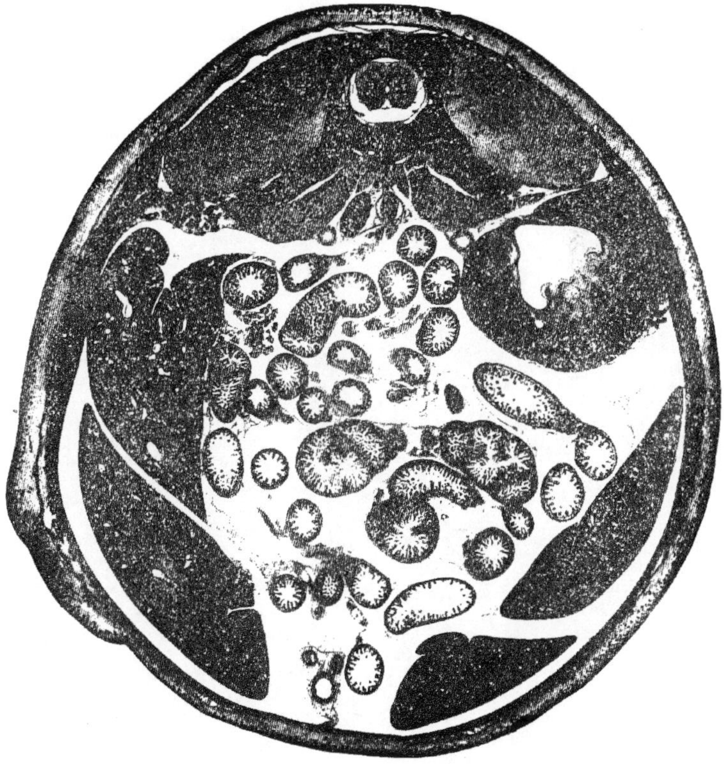

VIII.

A TRANSVERSE SECTION OF THE ABDOMEN

PART II.

OSTEOLOGY OF THE RABBIT.

For a practical study of the rabbit's skeleton, a thoroughly cleaned, but otherwise rough, unmounted skeleton will be found most convenient. The skull should be divided with a fine saw at a little to one side of the median plane, or a second skull may be provided for this purpose (cf. Fig. 62). The most useful specimens for reference are: (1) a well-mounted skeleton of the adult animal, showing the natural relations of the bones; and (2) a rough skeleton of a young animal of from one to five weeks, showing the primary composition of cartilage bones. For the special study of the skull (pp. 140-159) a disarticulated specimen may be employed, but the majority of the features may be made out in the intact or divided skulls. The general account of the skull as given below will be found to cover most of the osteological points noted in the dissection.

DIVISIONS OF THE SKELETON.

The skeleton is divisible into two main portions, namely, the **axial skeleton** and the **appendicular skeleton.** The former comprises the vertebral column, the ribs, the sternum, and the skeleton of the head; the latter, the supports of the anterior and posterior limbs, and the associated pectoral and pelvic girdles.

THE VERTEBRAL COLUMN.

The **vertebral column** (columna vertebralis) is formed of a linear series of segments, the **vertebrae.** In accordance with its function as a general support of the body, and also its relations with the nervous system and the spinal musculature, the vertebrae, with minor exceptions, are constructed on the same plan. Those of

particular regions also present certain features in common, so that it is possible to classify them into cervical, thoracic, lumbar, sacral, and caudal groups.

A typical vertebra—for the characters of which any one of the

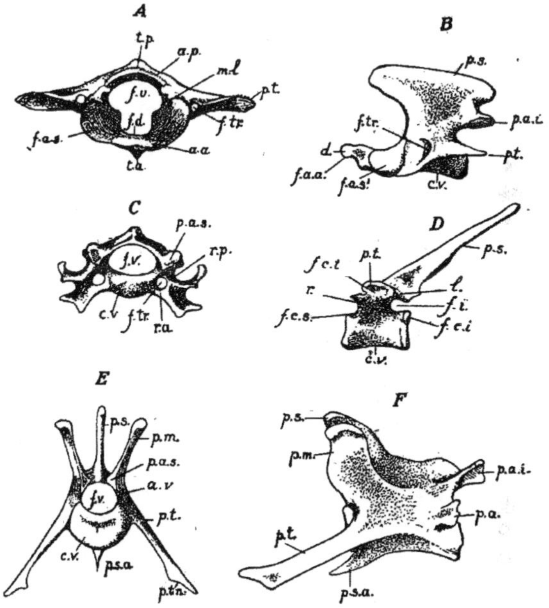

FIG. 56. Representative vertebrae: A, atlas, anterior surface; B, epistro pheus, lateral surface; C, fifth cervical vertebra, anterior surface; D, fourth dorsal, lateral surface; E, F, second lumbar vertebra, anterior and lateral surfaces.

a.a., anterior arch of atlas; a.p., posterior arch of atlas; a.v., vertebral arch; c.v., vertebral body; d., dens epistrophei; f.a.a., anterior articular facet of epistropheus; f.a.s., superior articular pit of atlas; f.a.s.', superior articular. facet of epistropheus; f.c.i., inferior costal demifacet for head of rib; f.c.s., superior costal demifacet; f.c.t., costal facet of transverse process; f.d., fovea dentis; f.i., intervertebral foramen; f.tr., foramen transversarium; f.v., foramen vertebrale; l., lamina of vertebral arch; m.l., lateral mass of atlas; p.a., accessory process of lumbar vertebra; p.a.i., inferior articular process; p.a.s., superior articular process; p.m., mamillary process; p.s., spinous process; p.s.a., anterior spinous process; p.t., transverse process; p.t., triangular process; r., radix of vertebral arch; r.a., r.p., anterior and posterior radices of transverse process of cervical vertebra; t.a., t.p., anterior and posterior tubercles of atlas.

thoracic or lumbar series may be taken (Fig. 56, D-F)—consists of a basal portion, the **vertebral body** (corpus vertebrae), and of a dorsal, **vertebral arch** (arcus vertebrae). The two portions

enclose a large aperture, the **vertebral foramen** (foramen vertebrale). The successive foramina form an almost complete tube, the **vertebral canal** (canalis vertebralis), for the accommodation of the spinal cord.

The body of a vertebra is a cylindrical, or somewhat dorsoventrally compressed, mass of bone, which bears at either end an articular surface for attachment to the adjacent vertebra. The articular surfaces are borne on thin plate-like epiphyses, the epiphysial lines being evident even in older animals, especially in the lumbar region. The dorsal portion of the body bears on either side the **pedicle,** or **root** of the vertebral arch (radix arcus vertebrae), the dorsal surface of the body forming in this way the floor of the vertebral foramen. The dorsal portion of the arch, borne on the pedicle, is distinguished as the **lamina.** The anterior and posterior margins of the pedicle are notched, each notch or incisure being converted, through its association with that of the adjacent vertebra, into a rounded aperture, the **intervertebral foramen** (foramen intervertebrale), for the passage outward of a spinal nerve.

The arch of the vertebra is noteworthy for its projections or **processes.** On either side is a horizontal plate of bone, the **transverse process** (processus transversus), and, dorsally, a median projection, the **spinous process** (processus spinosus), all three serving for the attachment of the vertebrae to one another by ligaments, and for the attachment of the spinal musculature. Special articular surfaces, borne on low **articular processes** (processus articulares), are found on the anterior and posterior margins of the arch. The anterior, or **superior articular surfaces** are directed for the most part toward the dorsal surface, and are overlapped in the natural condition by the **inferior articular surfaces,** which are directed toward the ventral surface. A certain amount of movement is permitted by one surface slipping across the other, the mechanism illustrating the **arthrodia,** or gliding-joint.

The **cervical vertebrae** (vertebrae cervicales) are seven in number. The posterior five are similar, while the anterior two are specially modified in relation to the skull. The posterior vertebrae (Fig. 56, C) are dorsoventrally compressed, their arches low, and the spinous process short. In the seventh vertebra, however, the spinous process begins to be elongated as in the succeeding thoracic

vertebrae. In each vertebra the transverse process is perforated by a **costo-transverse foramen** (foramen transversarium), the latter serving for the passage of the vertebral artery forward to the head. Through the presence of this aperture, the base of the transverse process is divided into two parts, namely, a dorsal, or **posterior root** (radix posterior), and a ventral, or **anterior root** (radix anterior). The anterior root is a coalesced rib, and is comparable in its general relations to the normal ribs of the thoracic vertebrae.

The first vertebra is the **atlas** (Fig. 56, A). It is peculiar in lacking the vertebral body, the latter being represented by the odontoid process of the epistropheus (cf. Plate II); also in possessing special articular surfaces, and in having its transverse process greatly flattened in the dorsoventral direction. It consists of a ventral half-ring, the **anterior arch** (arcus anterior), a dorsal half-ring, the **posterior arch** (arcus posterior), with paired **lateral masses** (massae laterales) uniting them. The lateral masses also form the bases of the transverse processes. The anterior arch bears on its ventral side a small backwardly-directed process, the **anterior tubercle** (tuberculum anterius). A similar **posterior tubercle** (tuberculum posterius) on the dorsal surface of the posterior arch is comparable to the spinous process of an ordinary vertebra. The anterior surface of the atlas bears on either side an extensive concave smooth surface, the **superior articular pit** (fovea articularis superior), for articulation with the convex occipital condyles of the skull. Its posterior surface bears on either side a somewhat triangular **inferior articular facet** (facies articularis inferior) for articulation with the epistropheus. These surfaces take the place of the arch articulations of ordinary vertebrae. Through the compression of the transverse process, the costo-transverse foramen is converted into a canal. The anterior aperture of this leads by a shallow groove, the **sulcus arteriae vertebralis**, into a foramen perforating the posterior arch. This aperture, represented in some mammals by separate alar and intervertebral foramina serves to transmit the vertebral artery and the first cervical nerve.

The space enclosed by the atlas is divided into a dorsal portion, corresponding to the vertebral foramen of other vertebrae, and a

ventral portion which in the natural condition lodges the odontoid process of the epistropheus. The division is effected partly by a small tubercle on the inner side of each lateral mass, and partly by a transverse ligament which is stretched between them and over the dorsal surface of the odontoid process. On the floor of the ventral portion, a rounded articular surface, the **fovea dentis,** marks the point of articulation of the anterior articular facet of the odontoid process with the inner surface of the anterior arch.

The second vertebra is the **epistropheus** or **axis** (Fig. 56, B). It resembles the succeeding cervical vertebrae more closely than does the atlas. It is noteworthy for its great size, for the lateral compression of its arch and spinous process, and for the possession of a stout forwardly-directed **odontoid process**, or **tooth** (dens epistrophei). It is articulated with the atlas through an **anterior articular facet,** borne on the ventral surface of the odontoid process, and by large paired **superior articular facets** borne on its base. The spinous process of this vertebra and the transverse processes of the atlas form together three main points of attachment for the occipital musculature.

The **thoracic vertebrae** (vertebrae thoracales) are twelve in number. They are distinguished chiefly by the possession of **articular pits** for the attachment of ribs (Fig. 56, D). A rib is articulated at two points, namely, one on the body of the vertebra, the other on the transverse process. The former is marked by a small round depression, the **costal pit** (fovea costalis), or **costal facet.** In the last two vertebrae the facet is borne wholly on the vertebral body to which the rib belongs. In the remaining vertebrae a complete articulating surface is formed by two **demifacets,** one being on the vertebra to which the rib belongs, the other on the vertebra immediately in front. The articulation of a rib with a transverse process is marked by an oval facet, the **costal pit of the transverse process** (fovea costalis transversalis). It is present only in the first ten of the thoracic vertebrae.

In all vertebrae of the thoracic series the spinous processes are well-developed. They increase in length to the third, and then become gradually shorter, although their surfaces, on the whole, are slightly increased in extent. The anterior ten are directed backward, the eleventh is almost vertical, while the twelfth is directed forward, like those of the succeeding lumbar vertebrae.

The **lumbar vertebrae** (vertebrae lumbales) are seven in number. They are large vertebrae, conspicuous for their extensive surfaces and processes for muscular attachment (Fig. 56, E, F). The transverse processes continue the general line of the ribs of the thoracic region. They are directed forward, as well as outward, and the tip of each is formed by a thin **triangular plate** (processus triangularis), which represents a fused rib. At the posterior side of the base of each is a short, flattened projection, the **accessory**

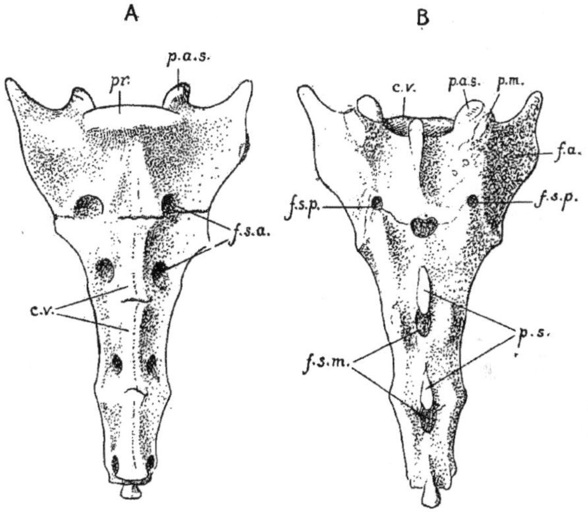

FIG. 57. The os sacrum: A, ventral (pelvic) surface; B, dorsal surface; c.v., bodies of coalesced vertebrae; f.a., auricular surface; f.s.a., anterior sacral foramina; f.s.m., median sacral foramina; f.s.p., posterior sacral foramina; p.a.s., superior articular process of first vertebra; p.m., mamillary process of first vertebra; pr., promontory; p.s., spinous processes.

process (processus accessorius). The spinous process is especially well-developed, and is directed forward. The articular processes are rotated upward, so that their surfaces are directed more nearly toward, or away from, the median plane, instead of to the dorsal or ventral surface. The anterior articular surfaces are borne on the bases of stout, upwardly-directed **mamillary processes** (processus mamillares). The latter are most characteristic of the lumbar vertebrae, but may be seen to arise in the posterior thoracic

region as small elevations of the transverse processes. Each of the first three of the lumbar vertebrae bears a median ventral projection, the **anterior spinous process** (processus spinosus anterior), for the attachment of the lumbar portion of the diaphragm.

The **sacral vertebrae** (vertebrae sacrales) are four in number. In contrast to the **true** vertebrae—those united by ligament and articular surfaces—of the remaining portions of the vertebral column, they are **false** vertebrae, united in the young by synchondroses, and in the adult coalesced to form a composite structure, the **os sacrum** (Fig. 57). The axis of the sacrum forms an obtuse angle with that of the lumbar vertebrae, the angle being indicated by a ventral projection, the **promontory** (promontorium), formed by the last lumbar and first sacral vertebrae. The sacrum is the medium through which the vertebral column—in other words, the posterior portion of the trunk—is supported on the posterior limbs. Its anterior dorsal portion bears on either side a roughened area, the **auricular surface** (facies auricularis), for articulation with the pelvic girdle. This surface is borne for the most part on the transverse process of the first sacral vertebra.

The sacrum exhibits many features resulting from its formation through the fusion of originally distinct vertebrae. On the ventral, or **pelvic surface** (facies pelvina), the lines of junction may be traced either between the bodies, or between the transverse processes. Four pairs of apertures on this surface, the **anterior sacral foramina** (foramina sacralia anteriora), lead into the intervertebral foramina, and give passage to the sacral spinal nerves. On the **dorsal surface** (facies dorsalis) a pair of **posterior sacral foramina** lie in the line of junction of the first and second vertebrae. The spinous processes are evident in all four vertebrae. The combined articular and mamillary processes are conspicuous only in the first two, but are represented in the succeeding two by low, roughened tubercles. In the middle line dorsally the vertebral arches are separated by conspicuous apertures, the **median sacral foramina.**

The **caudal** or **coccygeal vertebrae** (vertebrae caudales, s. coccygeae) are sixteen in number. There are segments of small size, increasing slightly to the third, and then gradually decreasing to the end of the column. The arches are complete in the first

seven. The transverse processes are vestigial in all except the third. At the end of the column the segments are reduced to slender cylinders of bone representing the vertebral bodies.

THE RIBS.

The **ribs** (costae) are twelve in number on either side. Each is composed of a dorsal portion, the **costal bone** (os costale), or **bone-rib,** and a ventral portion, the **costal cartilage** (cartilago costalis) (Fig. 58). From their attachment on the vertebral column the bone-ribs are directed outward, downward, and backward. The costal cartilages are directed for the most part inward, downward, and forward. The first costal cartilage forms a pronounced angle with the corresponding bone-rib. In the succeeding ribs the angle is gradually replaced by a broad curve.

Ribs are classified as **true** or **sternal ribs** (costae verae), and **false** or **asternal ribs** (costae spuriae). The former—comprising the anterior seven—are those directly attached to the sternum. The latter—comprising the posterior five—are either indirectly attached, or unattached. The unattached ribs are designated as **floating ribs.**

Generally speaking, the bone-ribs are cylindrical; but the anterior five or six are more or less flattened, with their main surfaces respectively medial and lateral. The compression is most marked in old animals. The first rib is extremely short. The succeeding ribs increase in length to the sixth, and then decrease to the twelfth. The arch formed by each rib has its greatest convexity, or **angle,** at some point toward the dorsal surface. Passing backward, the point of greatest convexity changes from a medial to a lateral position. This, together with the elongation of the more posterior ribs, results in an enormous increase in the posterior extent of the thoracic cavity.

The vertebral end, or **head** of the rib (capitulum costae), is articulated with the body of the vertebra to which it belongs, and also, in the case of the first ten, with the vertebra immediately in front. The articulation with a transverse process is marked by a small smooth elevation, the **costal tubercle** (tuberculum costae). It is present only in the first nine ribs. Except in the first rib, and

in the last four, the tubercle bears a sharp, dorsally-directed process for muscular attachment. The slender portion of the rib intervening between the head and tubercle is the **neck** (collum costae), the remaining larger portion being distinguished as the **body** of the rib (corpus costae).

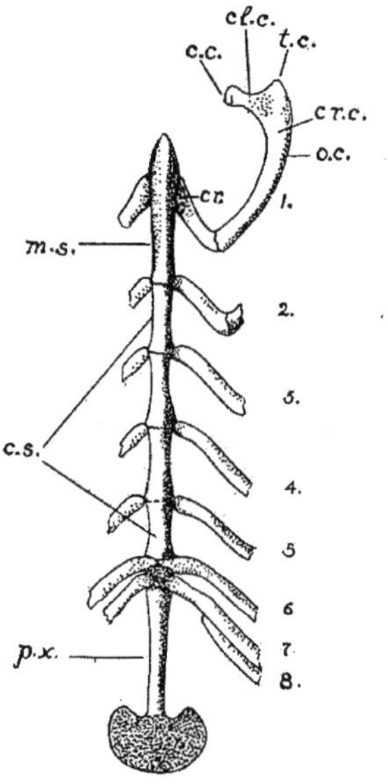

FIG. 58. The sternum and first rib, ventral view: 1-7, the true ribs; 8, first false rib; c.c., head of rib; cl.c., neck of rib; cr., costal cartilage; cr.c., body of rib; c.s., body of sternum; m.s., manubrium sterni; o.c., bone-rib; p.x., xiphoid process; t.c., costal tubercle.

The **bony thorax** is formed by the ribs and the sternum with the assistance of the thoracic vertebra. It encloses a large space, the **thoracic cavity** (cavum thoracis). The latter is conical in shape, with the apex directed forward. The dorsoventral diameter of the cavity is considerably greater than the transverse diameter. Apart from the **intercostal spaces**, the cavity is open at two points: anteriorly, the first thoracic vertebra, the first rib, and the manubrium sterni together enclose a small opening, the **superior thoracic aperture** (apertura thoracis superior); posteriorly, the seventh and succeeding ribs, together with the posterior thoracic vertebrae and the xiphoid process of the sternum, enclose a much larger opening, the **inferior thoracic aperture** (apertura thoracis inferior). In the natural condition it is largely closed by the diaphragm. The curved boundary formed by the ribs in this region is the **costal arch** (arcus costarum), the angle formed at the point of attachment of the xiphoid process being the **infrasternal angle** (angulus infrasternalis).

Through their articulations with the vertebral column, and the nature of the costal cartilages, the ribs are capable of being moved, or rotated, forward. The movement results in an increase of the extent of the thoracic cavity, and is of importance in respiration.

THE STERNUM.

The **sternum** (Fig. 58) consists of a linear series of six segments, the **sternebrae** The first segment is the **manubrium sterni.** It is about twice the length of the middle segments. It is somewhat triangular in section, two of its surfaces being ventrolateral, the third dorsal and directed toward the thoracic cavity. To its anterior tip is attached the **sternoclavicular ligament**, by which the greatly reduced clavicle is united with the sternum

The four middle segments are similar in appearance, and form the **body** (corpus sterni). The sixth segment, described as the **xiphoid process** (processus xiphoideus), is an elongated strip of bone, to the posterior end of which is attached a broad, thin plate of cartilage.

The first costal articulation is situated at about the middle of the manubrium, the remaining six at the points of junction of the segments. Five of them occur singly, while the sixth and seventh costal cartilages are attached together at the point of junction of the last segment of the corpus sterni with the xiphoid process.

THE SKELETON OF THE HEAD.

The head-skeleton comprises: (1) the series of elements constituting the **skull**; and (2) the **hyoid bone**, with its connections. The skull, or cranium—using that term in a general sense—includes the **cranium** proper, that portion enclosing the brain and containing in its wall the auditory capsules, and the **bones of the face** (ossa faciei), the latter including the series of elements related for the most part to the jaws and palate. The primary relations of the constituents of the head-skeleton have already been indicated above (p. 53).

A. THE SKULL AS A WHOLE.

The **skull** (Figs. 59-63) is a composite structure, consisting of a large number of elements, which, with the exception of the mandible, are united by synarthroses, so that they produce the effect of a continuous mass. The mandible is a more or less independent

structure, articulated with the main body of the skull by a typical joint.

The skull is roughly divisible into two portions, namely, an anterior, **facial portion,** and a posterior, **cranial portion** The cranial portion has a somewhat conical shape, its apex being directed forward. It is separated from the facial portion by a depression on either side of the skull, the **orbital cavity** (orbita), which serves for the accommodation of the optic bulb. Unlike the remaining special sense-organs, the eye is not included within the skull-wall. The two portions are united both medially and laterally, the lateral connection being established by the **zygomatic arch** (arcus zygomaticus), which bridges the lateral portion of the orbit. The facial portion has also a somewhat conical shape, its apex being formed by the anterior extremity of the upper jaw and the incisor teeth. Its base is formed in part by the connection with the cranial portion, as already described, and also by the anterior walls of the orbits.

The **cranial portion** exhibits an extensive posterior, **nuchal surface** (planum nuchale), situated in general at right angles to the cervical portion of the vertebral column and also to the dorsal, lateral, and ventral walls of the skull. This surface includes the external surface of the **occipital bone,** with the exception, chiefly, of the basilar portion of the latter. Its dorsal portion forms an area of attachment for the spinal and special occipital musculature. Its ventral portion is perforated by a large aperture, the **foramen magnum occipitale,** for the passage of the central nervous system from the cranial cavity into the vertebral canal. On either side of this is a smooth, ridge-like projection, the **occipital condyle** (condylus occipitalis), for articulation with the superior articular pits of the atlas. At a little distance lateral to the occipital condyle, the nuchal surface is continued downward through the medium of a somewhat triangular, pointed **jugular** or **paramastoid process** (processus jugularis). This structure is separated from the occipital condyle by a pronounced notch, the posterior boundary of a deep narrow excavation, the **jugular fossa** (fossa jugularis), which lies between the condyle and the tympanic bulla. The jugular process serves for the attachment of muscles belonging to the tongue, hyoid, and mandible, namely, the styloglossus, stylohyoidei

major and minor, and the digastricus, the suspensory ligament of the lesser cornu of the hyoid also being included in the ligament of the stylohyoideus minor. The nuchal surface is separated from the dorsal surface of the skull by a shield-shaped promontory and crest (crista nuchae). The lateral continuation of this crest is the **superior nuchal line** (linea nuchae superior). It forms a curved ridge, the position of which indicates the dorsal limit of the occipital musculature. The posterior, somewhat tri-radiate tip of the

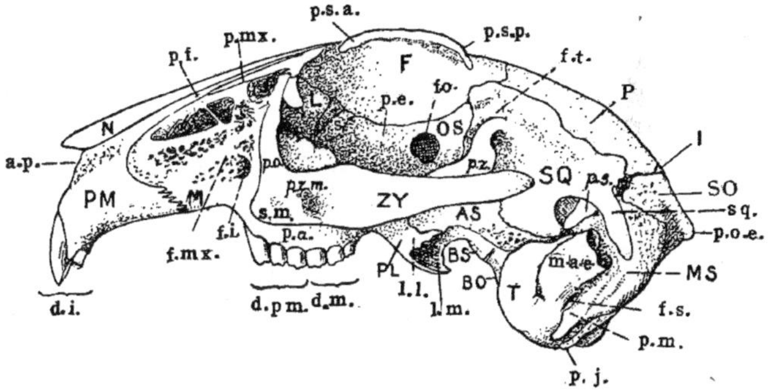

FIG. 59. Lateral surface of the skull: AS, alisphenoid (ala magna); BO, basioccipital (basilar portion of occipital); BS, basisphenoid (body of posterior sphenoid); F, frontal; I, interparietal; L, lacrimal; M, maxilla; MS, mastoid portion of petrosal (petromastoid); N, nasal; OS, orbitosphenoid (ala parva); P, parietal; PL, palatine; PM, premaxilla; SO, supraoccipital (squamous portion of occipital); SQ, squamosal; T, tympanic; ZY, zygomatic.

a.p., piriform aperture of nose; d.i., incisor teeth; d.m., molars; d.pm., premolars; f.i., infraorbital foramen; f.mx., maxillary fossa; f.o., optic foramen; f.s., stylomastoid foramen; f.t., temporal fossa; l.l., lateral lamina of pterygoid process; l.m., medial lamina; m.a.e., osseous portion of external acoustic meatus; p.a., alveolar process of maxilla; p.e., ethmoidal portion of orbitosphenoid; p.f., frontal process of premaxilla; p.j., jugular process of occipital; p.m., mastoid process of mastoid; p.mx., maxillary process of frontal; p.o., orbital process of maxilla; p.o.e., external occipital protuberance; p.s., squamosal process of parietal; p.s.a., and p.s.p., anterior and posterior supraorbital processes of frontal; p.z., zygomatic process of squamosal; p.z.m., zygomatic process of maxilla; s, sphenoorbital process of maxilla; s.m., spina masseterica; sq., squamosal process of squamosal.

projection, together with a thin ridge extending ventrad from it, is the **external occipital protuberance** (protuberantia occipitalis externa), an important median point of attachment for the occipital muscles and the ligamentum nuchae.

The ventral wall of the cranial portion is the basal portion (**basicranium**) of the entire skull. Its axial line, the **basicranial axis**, continues, in general, that of the bodies of the vertebrae, and its posterior portion is equivalent, morphologically, to vertebral

segments. It is formed by a linear series of three bones, namely, the **basilar portion** of the **occipital,** the **body** of the **posterior sphenoid,** and that of the **anterior sphenoid** (respectively **basi-occipital, basisphenoid** and **presphenoid bones**). Its extremely narrow, anterior portion forms the roof of a deep groove which encloses the nasal portion of the pharynx. As viewed from the ventral surface, it is seen to disappear in the facial complex at some distance dorsal to the posterior margin of the bony palatine bridge. Laterally, it is separated from the orbit on either side by a vertical plate formed by the palatine bone, and also by two downward projections of the posterior sphenoid, the **medial** and **lateral laminae** of the **pterygoid process** (processus pterygoideus). These structures enclose between them the **pterygoid fossa** (fossa pterygoidea), the walls of which serve for the attachment of the external and internal pterygoid muscles of the mandible.

The lateral wall of the cranial portion forms anteriorly a large part of the boundary of the orbit. The cranial wall of the orbit is partly formed by two upward projections of the basicranium, namely, the **lesser** or **orbital wing** of the **anterior sphenoid,** or **orbitosphenoid** and the **greater** or **temporal wing** of the **posterior sphenoid,** or **alisphenoid.** The remaining portion is formed by membrane elements, including the **frontal bone** of the roof of the skull and the **squamosal bone,** the latter distinguishable as the support of the posterior root of the zygomatic arch. Immediately behind the orbit, the root of the zygomatic arch projects outward and then downward. It is formed by a **zygomatic** process (processus zygomaticus) of the squamosal bone, and the tip of this forms a vertical plate, which is united by a horizontal suture with the zygomatic bone. On the ventral side of the process, close to the cranial wall, is the **glenoid cavity** or **mandibular fossa** (fossa mandibularis), for articulation with the mandible. On its dorsal side, but more especially on the adjacent portion of the cranial wall, there is a shallow, horizontal groove, lodging in the natural condition the temporalis muscle of the mandible, and therefore representing a greatly reduced **temporal fossa** (fossa temporalis). In the natural condition the anterior portion of the groove is converted into a foramen through the presence of a stout ligament extending from the posterior supraorbital process to the base of the zygomatic arch. The dorsal boundary of the fossa is formed by a pronounced

ridge, the **temporal line** (linea temporalis), the latter forming also the lateral margin of the roof of the skull in this region.

Behind the posterior root of the zygomatic arch, the external surface of the lateral wall is largely occupied by the swollen **tympanic bulla** (bulla tympani), formed by the tympanic bone. It contains the capacious **tympanic cavity** (cavum tympani) and certain skeletal structures of the middle ear, namely, the **auditory ossicles** (ossicula auditus), the relations of which are more fully dealt with below (p. 147). The dorsal portion of the tympanic bulla is continuous with a short bony tube which opens at a short distance dorsally by a large oval aperture. This tube is part of a more extensive canal, the **external acoustic meatus** (meatus acusticus externus) which, in the natural condition, leads downward through the base of the external ear to the tympanic membrane. The tympanic bulla does not form the lateral wall of the skull in this region, and is not exposed to the cranial cavity. It is applied closely to the external surface of the **periotic** or **petromastoid bone** (os petrosum), which forms the lateral boundary of the cranial cavity, and contains the structures of the internal ear. The external or **mastoid portion** of this bone appears in the space enclosed between the tympanic bulla and the jugular process of the occipital

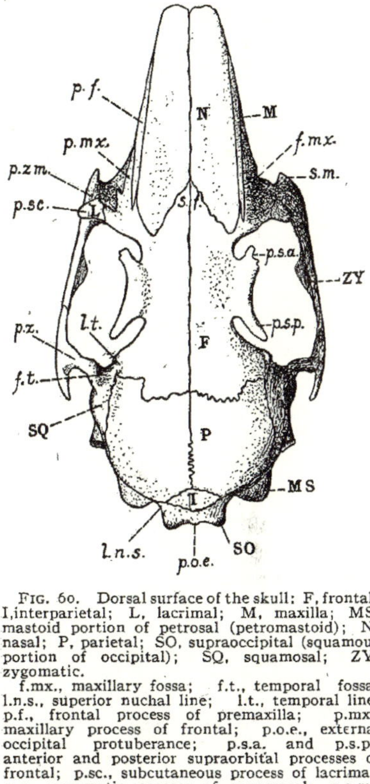

FIG. 60. Dorsal surface of the skull: F, frontal; I, interparietal; L, lacrimal; M, maxilla; MS, mastoid portion of petrosal (petromastoid); N, nasal; P, parietal; SO, supraoccipital (squamous portion of occipital); SQ, squamosal; ZY, zygomatic.

f.mx., maxillary fossa; f.t., temporal fossa; l.n.s., superior nuchal line; l.t., temporal line; p.f., frontal process of premaxilla; p.mx., maxillary process of frontal; p.o.e., external occipital protuberance; p.s.a. and p.s.p., anterior and posterior supraorbital processes of frontal; p.sc., subcutaneous process of lacrimal; p.z., zygomatic process of squamosal; p.z.m., zygomatic process of maxilla; s.f., frontal spine; s.m., spina masseterica.

bone, where it is readily distinguishable by its pitted appearance. Its ventral portion bears a slender projection, lying parallel to the jugular process, the **mastoid process** (processus mastoideus).

A series of foramina, lying partly within the orbit, and extending thence posteriorly along the boundary between the lateral and ventral walls to the occiput, put the cranial cavity in communication with the outside, and serve for the passage of nerves and vessels. The first and largest of these, the **optic foramen** (foramen opticum), occupies the middle portion of the orbit, and transmits, in the natural condition, the optic nerve. Following this is a vertical slit-like aperture—not to be confused with the perforations of the external lamina of the pterygoid process—the **superior orbital fissure** (fissura orbitalis superior). It represents both the superior orbital fissure of the normal mammalian skull and the **foramen rotundum**, and provides for the passage outward of the third, fourth and sixth cranial nerves, together with the first and second divisions of the fifth. The lateral lamina of the pterygoid process presents three foramina, of which the largest, anterior, and medial one, the **anterior sphenoidal foramen** (alar canal), serves for the transmission of the internal maxillary artery, while the remaining two, the **middle** and **posterior sphenoidal foramina**, transmit muscular branches (masseterico-temporal and pterygobuccinnator) of the mandibular nerve. On the medial side of the base of the medial lamina of the pterygoid process there is a shallow longitudinal groove, representing the **pterygoid canal** (canalis pterygoideus) of the human skull. Immediately in front of the tympanic bulla, on the ventral surface of the skull, an irregular aperture, the **foramen lacerum,** leads directly into the cranial cavity. It is incompletely divided into two parts by a slender bony splint. It contains, in addition to the foramen lacerum, the **foramen ovale** of the typical mammalian skull, and serves to transmit the mandibular portion of the fifth nerve and the internal carotid artery. Looking into the aperture from the front, it is seen to communicate not only with the cranial cavity, but also with two apertures in the anterior portion of the auditory complex. One of these—that toward the middle line—is the **internal carotid foramen** (foramen caroticum internum). It is the anterior end of a canal transmitting the internal carotid artery; the posterior end of this canal,

the point at which the internal carotid artery enters the tympanic bone, or the **external carotid foramen** (foramen caroticum externum), being visible as a rounded aperture lying on the posteromedial surface of the tympanic bulla. The second, lateral aperture communicating with the foramen lacerum is that of the **auditory** (Eustachian) **tube** (tuba auditiva). It leads into the tympanic cavity, and in the natural condition the tube places this cavity in communication with the nasal portion of the pharynx. Associated with the mastoid process is a small aperture, the **stylomastoid foramen** (foramen stylomastoideum), the external opening of the facial canal, which serves for the passage of the facial nerve. A slit-like aperture, the **jugular foramen** (foramen jugulare), lies in the jugular fossa, between the posterior ventral margin of the tympanic bulla and the occipital condyle. It transmits the first portion of the internal jugular vein from the transverse sinus of the dura mater, and also the ninth, tenth, and eleventh cranial nerves. Finally, immediately in front of the dorsal portion of the condyle, the occipital segment is perforated by several small apertures together representing the **hypoglossal canal** (canalis hypoglossi), and serving for the transmission of the hypoglossal nerve.

The roof of the cranial portion is largely formed by two pairs of thin membrane elements, the **frontal** and **parietal** bones. The former occupy a general position between the orbits, while the latter are interposed between the frontal bones and the occipital segment. A small portion of the roof is formed posteriorly, however, by a small lozenge-shaped element, the **interparietal** bone, and by the shield-shaped projection, described above, which is part of the occipital bone.

The space enclosed by the cranial portion of the skull is the **cranial cavity** (cavum cranii). Its form depends on the external configuration of the brain. It is divisible into three portions, known as the **cranial fossae.** The **anterior cranial fossa** (fossa cranii anterior) is a small division lodging in the natural condition the olfactory bulbs of the brain. The **middle cranial fossa,** the largest division of the cavity, lodges the enlarged cerebral hemispheres. The **posterior cranial fossa** is a small division extending backward to the foramen magnum and lodging in the

natural condition the cerebellum and related posterior portions of the brain. It is partly set off from the middle cranial fossa by a fold of the dura mater, the **tentorium cerebelli,** which projects inward from the dorsal and lateral walls of the skull. This fold is usually found adhering to the internal surface of the skull, unless the latter has been very thoroughly cleared, and in all cases its position is indicated by a low ridge of bone. The marked difference in diameter between the middle and posterior cranial fossae is accounted for by the great thickness of the auditory portion of the skull. The anterior surface of the periotic bone will be observed to form an extensive posterior wall for the middle cranial fossa.

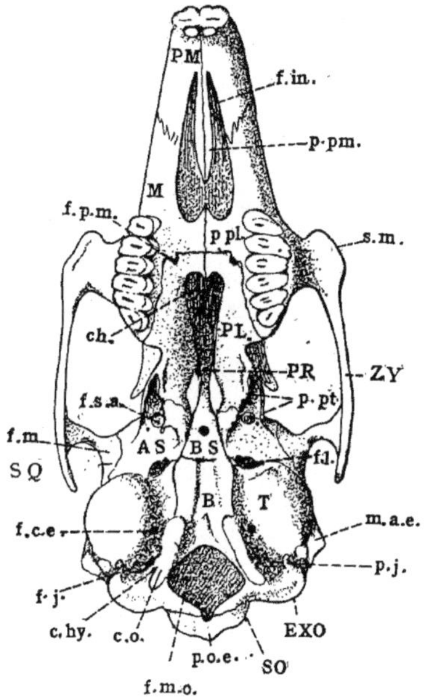

FIG. 61. Ventral surface of the skull: AS, alisphenoid (ala magna); B, basioccipital (basilar portion of occipital); BS, basisphenoid (body of posterior sphenoid); EXO, exoccipital; M, maxilla; PL, palatine; PMX, premaxilla; PR, presphenoid (body of anterior sphenoid); SO, supraoccipital (squamous portion of occipital); SQ, squamosal; T, tympanic; ZY, zygomatic.
ch, choana; c.hy., hypoglossal canal; c.o., occipital condyle; f.c.e., external carotid foramen; f.in., incisive foramen; f.j., jugular foramen; f.l., foramen lacerum; f.m., mandibular fossa; f.m.o., foramen magnum; f.p.m., greater palatine foramen; f.s.a., anterior sphenoidal foramen; m.a.e., osseous portion of external acoustic meatus; p.j., jugular process; p.o.e., external occipital protuberance; p.pl., palatine process of maxilla; p.pm., palatine process of premaxilla; p.pt., medial and lateral laminae of pterygoid process of posterior sphenoid; s.m., spina masseterica.

The floor of the middle and posterior cranial fossae is not smooth, like the external base of the skull, but presents in its anterior portion a prominent elevation, the **sella turcica**, which is borne on the body of the posterior sphenoid. It contains a large central depression, the **hypophyseal fossa** (fossa hypophseos), which in the natural condition lodges the hypophysis or pituitary body. The aperture of the fossa is partly enclosed laterally by a pair of pointed **posterior clinoid processes** (processus clinoidei posteriores), the

tips of which are directed forward; and a corresponding pair of **anterior clinoid processes** lie at the anterior end of the fossa, with the tips directed backward. The posterior, and also dorsal wall of the fossa, described as the **dorsum sellae,** leads by an abrupt curve backward on to the floor of the posterior cranial fossa, the sloping portion of the floor, or **clivus,** supporting in the natural condition the pons and medulla oblogata. Toward the anterior end of the middle cranial fossa, the lateral walls of the skull are greatly compressed, so that the anterior portion of the basicranium, especially the body of the anterior sphenoid, is largely ecxluded from the cranial cavity. The usually paired optic foramina are here confluent, there being a single aperture for the transmission of the optic nerves. The posterior ventral boundary of this aperture contains a broad groove, the **sulcus chiasmatis,** which lodges in the natural condition the optic chiasma.

In the anterior cranial fossa the floor is largely formed by a perforated area, borne on the **cribriform plate** (lamina cribrosa) of the **ethmoid bone,** and serving for the transmission of the divided olfactory nerves. Its median portion projects slightly into the cranial fossa as a low ridge, the **crista galli,** which is interposed between the tips of the olfactory bulbs.

In the ventrolateral portion of the cranial cavity may be found the internal openings of the foramina described above, namely, the superior orbital fissure, the foramen lacerum, the jugular foramen, and the hypoglossal canal. The superior orbital fissure is almost ventral in position to the foramen opticum, and is connected backward with the foramen lacerum by a broad groove, the **sulcus sphenoidalis,** which lodges in the natural condition the roots of the fifth nerve. This groove continues to the medial surface of the periotic bone, where it is bridged over by the tentorium cerebelli.

On the lateral wall of the posterior cranial fossa, and enclosed by the compact, white, **petrous portion** of the **peroitic bone,** is a series of three apertures leading into its substance. One of these, much larger than the remaining two, is the **parafloccular fossa** (fossa parafloccularis). It lodges in the natural condition the flocculus, a small stalked appendage of the cerebellum. Ventral of this fossa, and also somewhat in front of it, a thin lodge of bone extends over an oval opening, the internal

aperture of the **facial canal** (canalis facialis), which serves for the transmission of the seventh cranial (facial) nerve. Immediately behind and below this aperture is the opening of the **internal acoustic meatus** (meatus acusticus internus) for the transmission of the eighth cranial (acoustic) nerve. The two apertures tend to be enclosed by a shallow bony ridge, largely formed by the projecting ledge described above, and resembling superficially the complete common tube represented by the internal acoustic meatus of the human skull.

The **facial portion** of the skull is largely formed by the investing bones of the upper jaw, palate, and mandible, but it encloses also the entire olfactory region of the primary skull, including the nasal fossae and associated turbinal bones. The upper jaw—the **maxilla** of the human skull—is formed of two primary, and, in the rabbit, separate, elements, the **maxilla** and **premaxilla.** They together form the greater portion of the facial region—in the adult condition also a large portion of the lateral walls of the nasal fossae—and bear in a ventrolateral position low **alveolar processes** (processus alveolares) for the sockets or alveoli of the incisor and cheek-teeth. The maxilla bears the anterior root of the zygomatic arch, the latter being formed partly by a short **zygomatic process** arising from its lateral surface, by the **zygomatic** or **malar bone,** which is fused with it, and by the corresponding zygomatic process of the squamosal bone, constituting the posterior root. The anterior root of the zygomatic arch is perforated by a deep narrow **infraorbital canal** (canalis infraorbitalis), which opens on the facial surface by a vertical slit-like aperture, the **infraorbital foramen.** It serves for the transmission of the infraorbital vessels and nerves from the orbit to the face.

The ventral portion of the maxilla is associated with the **palatine bone** to form the **hard palate** (palatum durum). This structure is represented chiefly by a bony **palatine bridge** connecting the two sides of the skull between the more anterior cheek-teeth. It forms a portion of the roof of the oral cavity and a portion of the floor of the nasal cavity. Immediately in front of it, the palatal surface is perforated by a pair of large **incisive foramina** (foramina incisiva), which are broadly open to the nasal fossae.

A considerable portion of the anterior and dorsal wall of the orbit is formed by the facial complex. Dorsally, the roofing element of this region, the **frontal bone,** bears a curved lateral projection, the **supraorbital process** (processus supraorbitalis), which overhangs the orbit. Its narrower base expands into anterior and posterior tips, which lie parallel to the adjacent portion of the skull, and enclose with the latter corresponding **anterior** and **posterior supraorbital incisures.** The latter are converted by ligament into foramina. The anterior wall of the orbit is formed in part by a loosely articulated element, the **lacrimal bone,** the lateral margin of which projects from the orbital rim as a blunt **subcutaneous process** (processus subcutaneus). On the ventral side of its base is the orbital opening of the **nasolacrimal canal** (canalis nasolacrimalis), the bony enclosure of the nasolacrimal duct, which in the natural condition leads from the corneal surface of the eye to the anterior portion of the nasal fossa. A smaller projection forming the ventral boundary of the nasolacrimal aperture is the **hamulus lacrimalis.** Finally, in the ventral anterior angle of the orbit, the bases of the three posterior cheek-teeth encroach to a considerable extent on the orbital space. They are separated from the orbital wall by a deep **infraorbital groove** (sulcus infraorbitalis), which leads forward into the canal of the same name. They partly conceal two important apertures of this region, the orbital opening of the **pterygopalatine canal** (canalis pterygopalatinus), leading to the palatal surface, and the **spheno-palatine foramen** (foramen sphenopalatinum), leading to the nasal fossa. The pterygopalatine canal opens ventrally in the palato-maxillary suture of the hard palate by a rounded aperture, the **greater palatine foramen** (foramen palatinum majus).

The **nasal cavity** (cavum nasi) is enclosed by the maxilla and premaxilla, with the assistance of paired roofing elements, the **nasal bones.** Apart from the incisive foramina, which are closed in the natural condition, the cavity is open at two points. Posteriorly it communicates with the ventral surface of the skull by the **choanae,** which, in the rabbit, are incompletely divided. Anteriorly it opens to the outside by the **piriform aperture** (apertura piriformis). The cavity is divided into right and left portions, the **nasal fossae.** In the divided skull it is seen that the division is effected chiefly through a median vertical, cartilaginous plate,

the **nasal septum** (septum nasi), or cartilaginous portion of the **mesethmoid.** This is continuous posteriorly with a small crescentic vertical plate of bone, the **perpendicular plate** (lamina perpendicularis) of the ethmoid bone—the bony portion of the mesethmoid—and the latter is also the terminal element of the series of median bones constituting the basicranium. Posteriorly, the ventral portion of the cartilaginous nasal septum is supported by a vertical bony plate, the **vomer,** the dorsal margin of which is grooved to receive it. Anteriorly, the nasal septum bears on its ventral margin the paired enclosures of the vomeronasal organ,

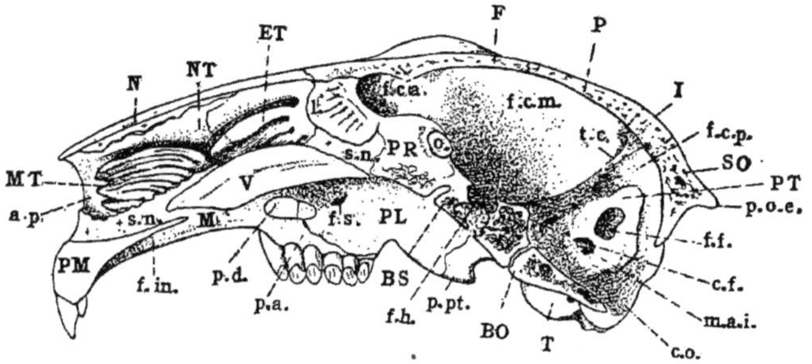

FIG. 62. The skull in vertical section: BO, basioccipital (basilar portion of occipital); BS, basisphenoid (body of posterior sphenoid); ET, ethmoturbinal; F, frontal; I, interparietal; M, maxilla; MT, maxilloturbinal; N, nasal; NT, nasoturbinal; P, parietal; PL, palatine; PMX, premaxilla; PR, presphenoid (body of anterior sphenoid); PT, petrous portion of petromastoid; SO, supraoccipital (squamous portion of occipital); T, tympanic; V, vomer.

a.p., piriform aperture of nose; c.f., internal aperture of facial canal; c.o., occipital condyle; f.c.a., f.c.m. and f.c.p., anterior, middle, and posterior cranial fossae; f.f., parafloccular fossa; f.h., hypophyseal fossa; f.in., incisive foramen; f.s., sphenopalatine foramen; l., perpendicular plate of the ethmoid; m.a.i., internal acoustic meatus; o., optic foramen; p.a., alveolar process of maxilla; p.d., hard palate; p.o.e., external occipital protuberance; p.pt., pterygoid process of posterior sphenoid; s.n., nasal septum; t.c., tentorium cerebelli.

which are also supported by the grooved surface formed in the middle line by the adjacent dorsal surfaces of the palatine processes of the premaxilla. The relations of these structures, as well as of the cartilage supporting the nasopalatine duct, are best seen in very young animals (cf. Plate III).

The delicate, folded, or scroll-like **turbinated bones,** characteristic of the nasal cavity, are borne on its posterior and lateral walls. Occupying the anterior portion of the lateral wall of the nasal fossa is a finely-ridged mass of bone, the **concha inferior,** or **maxilloturbinal.** It is easily distinguishable from a more dorsal

and posterior series of broader folds, which together constitute the **ethmoturbinal.** In the rabbit, as in mammals generally, the latter is divisible into a more dorsal elongated portion attached to the nasal bone, the **nasoturbinal,** and a more ventral portion, also posterior portion, the **ethmoturbinal proper,** composed of several shorter folds decreasing in length from above downward. In the natural condition the turbinated bones bear a considerable portion of the nasal epithelium, the surface of which is greatly increased by the folding of the underlying bone. That covering the ethmoturbinal contains the olfactory sense organs, while that covering the maxilloturbinal is non-sensory and possesses the

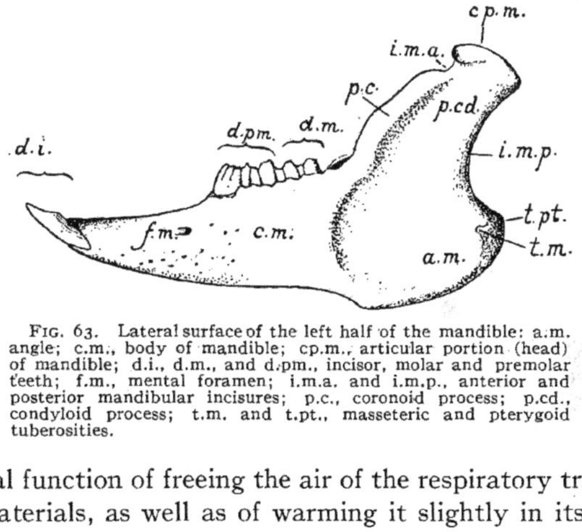

FIG. 63. Lateral surface of the left half of the mandible: a.m. angle; c.m., body of mandible; cp.m., articular portion (head) of mandible; d.i., d.m., and d.pm., incisor, molar and premolar teeth; f.m., mental foramen; i.m.a. and i.m.p., anterior and posterior mandibular incisures; p.c., coronoid process; p.cd., condyloid process; t.m. and t.pt., masseteric and pterygoid tuberosities.

mechanical function of freeing the air of the respiratory tracts from foreign materials, as well as of warming it slightly in its passage. On this account the respective structures are conveniently distinguished as **sensory (olfactory) turbinals** and **respiratory turbinals.**

The **mandible** (mandibula) is composed of two portions, united anteriorly by the **symphysis mandibulae.** Each half comprises a horizontal portion, forming in conjunction with that of the opposite side the **body** of the mandible (corpus mandibulae), and a posterior, vertical portion, the **ramus mandibulae,** the latter serving for the insertion of the muscles of mastication and for articulation with the skull. The body of the mandible bears on its

dorsal margin the alveoli of the lower teeth. The mandibular ramus forms a broad plate, the lateral surface of which is occupied in the natural condition by the masseter muscle, while the medial surface forms an area of insertion for the external and internal pterygoids. The surface of the ramus is greatly increased in its posteroventral portion through the expansion of the bone to form the **angle** (angulus mandibulae), or **angular apophysis.** The elongated articular surface is borne at the end of a vertical, or slightly oblique **condyloid process** (processus condyloideus). The nerve and vessels of the mandible enter at the **mandibular foramen** (foramen mandibulare), the latter being situated on the medial surface of the bone immediately behind the last cheek-tooth.

B. THE BONES OF THE SKULL.

1. THE OCCIPITAL BONE.

The **occipital bone** (os occipitale) is the first of the basicranial segments as numbered from the occipital articulation forward. It forms the posterior boundary of the skull, and establishes the connection of the latter with the vertebral column. Its external surface is identifiable for the most part with the nuchal surface, but a portion of it falls in the horizontal plane of the basis cranii. The internal surface is partly exposed to the cranial cavity, and forms in this relation the posterior, dorsal, and ventral boundaries of the posterior cranial fossa. The remaining portion is excluded from the cranial cavity, being applied instead to the broad posterior surfaces of the petrotympanic bones.

The occipital bone is divisible into four portions, namely, the **basilar portion** (pars basilaris) or **basioccipital,** the paired **lateral portions** (partes laterales), or **exoccipitals,** and the **squamous portion** (squama occipitalis), or **supraoccipital.** All four portions take part in the formation of the foramen magnum. In the young animal (Fig. 12) they are represented by separate elements, formed in a continuous mass of cartilage, and united for a time by synchondroses, but in the course of growth they become fused to form a single occipital bone.

The **basioccipital** is that portion lying below and in front of the foramen magnum. Its main surfaces are respectively dorsal and ventral. Its anterior margin is united with the posterior margin of the basisphenoid by a thin, transverse cartilage union, the **sphenooccipital synchondrosis** (synchondrosis spheno-occipitalis). Posteriorly its dorsal and ventral surfaces come together in a thin concave edge which forms the ventral boundary of the foramen magnum. Laterally it is bounded by the petro-tympanic bone and by the lateral portion of the occipital. The dorsal surface bears a median groove, deeper in its middle portion, where the lateral margins of the bone are raised to form a pair of rounded bosses for articulation with the petrotympanic. The groove represents the sloping portion or **clivus** of the occipital, and lodges in the natural condition, as described above, the ventral portion of the medulla oblongata. The ventral surface presents a similar groove, in the posterior portion of which there is a small ridge-like elevation, the **pharyngeal tubercle** (tuberculum pharyngeum).

The **exoccipital** is directed dorsad from the basioccipital in such a way that it falls in the plane of the nuchal surface. It is applied to the posterior surface of the petrotympanic bone, and also extends downward beyond the latter as the **jugular process.** The occipital condyle is borne on the exoccipital, with the exception, however, of its ventral tip, which belongs to the basioccipital. The portion of the occipital bone connecting the basioccipital and exoccipital contains the **jugular fossa** and the apertures representing the **hypoglossal canal.** Its anterior margin bears a **jugular incisure** (incisura jugularis), forming the occipital boundary of the jugular foramen, the remaining portion of the latter being formed by the petrotympanic.

The **supraoccipital** is the dorsal portion of the bone. Its dorsal margin is bent sharply forward, so that it tends to fall, like the basioccipital, in a horizontal plane. Its external surface bears the **nuchal crest** and the **external occipital protuberance.** A pair of lateral wing-like expansions rest upon, and partly overlap, the dorsal margins of the petrotympanic bones. The anterior boundary is formed by the interparietal, parietal, and squamosal bones, but in young skulls the squamosal connection is represented

by a vacuity. The internal surface bears a longitudinal groove, lodging in the natural condition the median vermis of the cerebellum. It is crossed at its anterior end by a shallow **transverse groove** (sulcus transversus), which marks the position of the transverse sinus of the dura mater.

2. THE POSTERIOR SPHENOID.

The **sphenoid bone,** as identified from the human condition, is a complex of elements belonging to two segments, namely, the **posterior sphenoid** (os sphenoidale posterius) and the **anterior sphenoid** (os sphenoidale anterius). In the rabbit, as in mammals generally, these segments are separate throughout life.

The **posterior sphenoid** comprises: (1) a median portion, the **body,** or **basisphenoid;** (2) paired dorsolateral expansions, the **greater wings** (alae magnae), or **alisphenoids;** and (3) paired ventral projections, the **pterygoid processes.**

The **basisphenoid** continues the basis cranii forward from the basioccipital to the body of the anterior sphenoid. It is united with the latter by the **intersphenoidal synchondrosis.** Its surfaces correspond for the most part to those of the basioccipital. The ventral surface forms the chief part of the bony roof of the nasopharynx. It is perforated in its middle by a round aperture, the **foramen cavernosum,** which leads into the interior of the bone. The dorsal surface is occupied by the **hypophyseal fossa** and related structures, namely, the **dorsum sellae** and the **posterior clinoid processes.** On the lateral surface of the base of the posterior clinoid process a faint groove, the **sulcus caroticus,** marks the course of the internal carotid artery. The interior of the bone contains a cavity of considerable size, the **sphenoidal sinus** (sinus sphenoidalis), which communicates both with the foramen cavernosum and the hypophyseal fossa.

The **alisphenoid** extends at first laterad, but soon changes its direction so that its axis beomes dorsoventral. At the same time the bone is rotated in such a way that its surfaces tend to fall in a transverse plane. It is bounded anteriorly by the orbitosphenoid, dorsally by the squamosal, and posteriorly by the petrotympanic. The anterior margin of its root encloses with the basisphenoid,

and to a certain extent with the orbitosphenoid, the **superior orbital fissure.** The **foramen lacerum** is formed by the posterior margin of its root in association with the petrotympanic.

The external surface of the alisphenoid is convex, both toward the orbit and toward the ventral surface of the skull. In the posterior portion of the orbit this surface bears a jagged elevation, the **crista alae magnae**. The internal surface forms a portion of the floor and anteroventral wall of the middle cranial fossa. At its base a broad groove, the **sulcus sphenoidalis,** indicates the position of the root of the fifth nerve and the related semilunar (Gasserian) ganglion.

The **pterygoid process** comprises the two plates described above as the **medial** and **lateral laminae.** The former is vertical, and its medial surface is directed toward the nasopharynx. The latter is almost horizontal. The medial lamina ends ventrally in a hooked projection, the **hamular process** (hamulus pterygoideus). In the young animal this portion is formed of an elevation of cartilage tipped by a separate membrane element, the pterygoid bone. The **pterygoid fossa** is formed in part by the medial and lateral laminae and in part by the divided posterior end of the palatine bone. The posterior basal portion of the lateral lamina is extensively excavated, like the adjacent portions of the alisphenoid. It bears a shallow groove, representing a **pterygoid canal** (canalis pterygoideus), and is perforated by the three apertures described above as the **anterior, middle,** and **posterior sphenoidal foramina.**

3. THE ANTERIOR SPHENOID.

The **anterior sphenoid** (os sphenoidale anterius) consists of two portions, namely, a median portion, the **body,** or **presphenoid,** and a pair of lateral expansions, the **lesser wings** (alae parvae), or **orbitosphenoids.**

The **presphenoid** is a constricted bony splint which continues the basis cranii forward from the basisphenoid. It is joined anteriorly with the perpendicular plate of the ethmoid and with the cartilaginous nasal septum. In the divided skull, or better in one from which the roof has been removed, the actual dorsal surface

of the bone is seen to be exposed to the cranial cavity only in its posterior portion, where it is occupied by the **sulcus chiasmatis** and the **optic foramina.** That part of the floor immediately in front of the optic foramina is formed by the coalesced roots of the orbitosphenoids, the dorsal surface of the presphenoid being thus excluded.

The **orbitosphenoid** forms a long, low plate, lying in the ventral portion of the orbit, and divided by a shallow notch at the level of the optic foramen into a posterior portion, the orbitosphenoid proper, and an anterior portion, the **ethmoidal process** (processus ethmoidalis). The orbitosphenoid proper lies behind the optic foramen. It is in contact dorsally with the orbital portion of the frontal, and ventrally with the alisphenoid; it assists the latter in the formation of the superior orbital fissure. Its posterior tip is in contact with the squamosal. Its internal surface forms a considerable portion of the anteroventral wall of the middle cranial fossa.

The **ethmoidal process** extends forward from the optic foramen. Its dorsal margin is articulated with the orbital portion of the frontal, and its ventral margin with the orbital portion of the palatine. Anteriorly it projects toward the lacrimal bone, thus occupying, in part, a space which, in the typical mammalian skull, is filled by the lamina papyracea of the ethmoid. Its internal surface is associated with the ethmoid bone and with the nasal cavity. It falls for the most part below the level of the cranial cavity.

4. THE SQUAMOSAL BONE.

The **temporal bone,** or **temporal complex,** as recognized from the human condition, is an association of three elements—**squamosal, tympanic,** and **periotic**—which in the human skull are coalesced to form a single bone. It is usually described as consisting of four portions, of which the **squamosal** and **tympanic portions** are two, while the periotic bone is considered to consist of two others, one of which, the **petrous portion,** is a solid white portion lodging the internal ear, while the second, or **mastoid portion,** is a mass of less compact character appearing externally

in the wall of the skull. In the rabbit the original elements are not coalesced, but the periotic and tympanic bones are so closely associated that it is proper to describe them as forming a **petrotympanic bone.**

The **squamosal bone** (os squamosum) is a rectangular plate, forming part of the lateral wall of the cranium, and bearing the posterior root of the zygomatic arch. It is articulated anteriorly with the orbitosphenoid and with the orbital portion of the frontal, dorsally with the frontal and parietal, posteriorly with the supraoccipital and petrotympanic, and ventrally with the alisphenoid. Its posterior margin bears a prominent, slightly decurved **squamosal process** (processus squamosus). It lies on the lateral surface of the petrotympanic immediately above the opening of the bony external acoustic meatus. The posterior root of the zygomatic arch is formed by a lateral and afterwards ventral projection, the **zygomatic process** of the squamosal. Its base bears ventrally the **mandibular fossa,** and dorsally, in association with the body of the squamosal, the **temporal fossa.** The internal surface of the squamosal forms a considerable portion of the wall of the cranial cavity, the middle cranial cavity being, in fact, broadest in this region.

5. THE PETROTYMPANIC BONE.

The **petrotympanic bone** (os petrotympanicum) is a somewhat oblong bone lying in the lateral wall of the cranium between the posterior sphenoid and occipital bones. It is chiefly indicated externally by the **tympanic bulla** and the bony **external acoustic meatus.** It is articulated anteriorly with the alisphenoid and squamosal, dorsally with the supraoccipital, and posteriorly with the exoccipital. Except for the presence of the squamosal process of the squamosal bone, the lateral and ventral surfaces are exposed to the outside of the skull. The internal surface is exposed to the posterior cranial fossa, with the exception, however, of a small ventral portion which is articulated with the basioccipital bone. Only a small portion of the anterior surface is in contact with the squamosal bone, the larger part being applied to the tentorium cerebelli and forming with the latter a posterior wall for the middle

cranial fossa. The dorsal portion of the bone corresponds in thickness with the wing of the supraoccipital with which it is articulated. The posterior surface is applied to the anterior surface of the exoccipital, and is thus excluded both from the cranial cavity and from the external surface of the skull. Viewing the skull from behind, however, it is seen that a small dorsal portion protrudes in a triangular space formed by the dorsolateral margin of the exoccipital and the ventrolateral margin of the supraoccipital wing. This portion is distinguishable by its pitted character. It forms the **mastoid portion** (pars mastoidea) as distinguished from the solid white **petrous portion** (pars petrosa), which is exposed to the cranial cavity, and which contains the structures of the internal ear. The mastoid portion lies for the most part above the tympanic cavity, but it is also continued ventrad between the external acoustic meatus and the exoccipital as the **mastoid process.** The **stylomastoid foramen** lies between the latter and the external acoustic meatus.

The petrous portion, as viewed from its medial surface, is roughly oblong; it is placed obliquely with reference to the basioccipital and basisphenoid. The **parafloccular fossa** occupies its posterodorsal portion, and extends into the substance of the bone, forming a much larger depression than is indicated by the diameter of its rim. The related dorsal margin of the bone is occupied by a groove which leads into a canal at its posterior margin. It indicates the position of the lateral portion of the transverse sinus of the dura mater. The ventral, thicker portion of the bone, enclosing the apertures of the **internal acoustic meatus** and the **facial canal,** is also that lodging the vestibulum and cochlea of the internal ear. A small aperture at its anteroventral angle, only visible when the petrotympanic is freed from its connections, represents the **hiatus canalis facialis** of the human skull. It transmits the great superficial petrosal nerve, a branch of the facial nerve passing to the sphenopalatine ganglion.

The tympanic surface of the petrous portion is described below in connection with the structures of the tympanic cavity.

The **tympanic portion** forms the spherical, expanded, shell-like, tympanic bulla, which contains in its interior the **tympanic cavity,** and is continuous dorsally with the bony enclosure of the

external acoustic meatus. The boundary between the two is indicated externally by a shallow oblique groove, the position of which indicates roughly that of the tympanic membrane within. The medioventral margin of the bone is articulated with the basioccipital, but the swollen portion is separated from the latter by a broad groove terminating posteriorly in the jugular fossa and the jugular foramen. Immediately in front of the jugular fossa, the rounded aperture of the **external carotid foramen,** transmitting in the natural condition the internal carotid artery, leads into the **carotid canal** of the interior of tympanic portion. At the anterior end of the groove, communicating with the foramen lacerum, is the anterior opening of the carotid canal, the **internal carotid foramen,** and on its lateral side the much larger aperture of the **auditory** (Eustachian) **tube.** The relations of these apertures are seen to best advantage when the petrotympanic is disarticulated from the associated posterior sphenoid bone. The auditory tube is then seen to lead directly into the tympanic cavity. A fine bristle may be passed through the carotid canal from one foramen to the other.

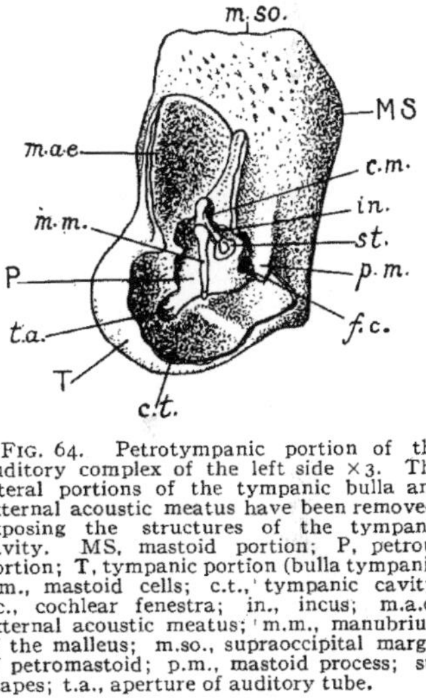

FIG. 64. Petrotympanic portion of the auditory complex of the left side ×3. The lateral portions of the tympanic bulla and external acoustic meatus have been removed, exposing the structures of the tympanic cavity. MS, mastoid portion; P, petrous portion; T, tympanic portion (bulla tympani). c.m., mastoid cells; c.t., tympanic cavity; f.c., cochlear fenestra; in., incus; m.a.e., external acoustic meatus; m.m., manubrium of the malleus; m.so., supraoccipital margin of petromastoid; p.m., mastoid process; st., stapes; t.a., aperture of auditory tube.

6. THE STRUCTURES OF THE TYMPANIC CAVITY

The relations of the tympanic cavity and associated structures may be studied with advantage in a skull from which the lateral wall of the tympanic bulla and external acoustic meatus has been removed, the surface displayed being as indicated in Fig. 64. The **tympanum** or **middle ear** is enclosed by the tympanic and petromastoid portions of the temporal complex. The attached

margin of the tympanic bulla encloses a roughly triangular area, into the ventral part of which the petrous portion of the petromastoid projects as a smooth, white, convex ridge, the **promontory** (promontorium). Above and behind the promontory the tympanic cavity is extended toward the mastoid portion of the bone as the **tympanic** or **mastoid antrum** (antrum tympanicum), and the interior of the mastoid portion is partly occupied by small extensions of the tympanic antrum, termed the **mastoid cells** (cellulae mastoideae). At the anteroventral angle of the area already described, a deep notch indicates the point of entrance of the auditory tube. The exposed surface of the petromastoid presents two apertures, one of which, situated posteroventrally, is open in the dried skull, and is the **cochlear fenestra** (fenestra cochleae). In the natural condition it is closed by a thin membrane which separates the tympanic cavity from the perilymphatic space containing the membranous labyrinth. The second aperture, the **vestibular fenestra** (fenestra vestibuli), lies above and in front of that just described. It is closed by the base of the stapes.

The **auditory ossicles** (ossicula auditus) comprise three elements, namely, the **malleus, incus,** and **stapes,** which bridge the space intervening between the tympanic membrane and the opening to the internal ear as represented by the vestibular fenestra. They occupy the dorsal angle of the triangular area already described, and lie immediately above the promontory. The malleus is the lateral element. The main portion, termed the **head,** is concealed by the projecting edge of the external acoustic meatus. It bears a stout vertical process, the **manubrium mallei,** which in the natural condition lies in contact with the tympanic membrane. The incus is the intermediate element; it is directly articulated with the malleus, and bears a downwardly-directed **long limb** (crus longum), for articulation with the minute head of the stapes. The latter element is a small stirrup-shaped bone, occupying an almost transverse position, and articulated at its base with the margin of the vestibular fenestra.

7. THE INTERPARIETAL BONE.

The **interparietal** (os interparietale) is a small, lozenge-shaped element, surrounded by the two parietal bones and the supraoccipital. It is the first of the membrane roofing elements of the cranium proceeding forward from the supraoccipital, and in the rabbit's skull is not fused with the occipital segment.

8. THE PARIETAL BONE.

The **parietal bone** (os parietale) is a characteristic roofing bone covering a large portion of the middle cranial fossa. It is somewhat rectangular in shape, and is connected by serrate sutures with the surrounding elements and with its fellow of the opposite side, the sutures producing a characteristic pattern on the external surface of the skull. The sutures are medial, anterior, lateral, and posterior in position, and are designated respectively as **saggital**, **coronal**, **squamosal**, and **lambdoidal.** The posterolateral angle of the bone is produced ventrally into a long, curved **squamous process** (processus squamosus), which lies in the angle formed by the tentorium cerebelli and the lateral wall of the middle cranial fossa. It is not exposed to the external surface of the skull.

9. THE FRONTAL BONE.

The **frontal bone** (os frontale) is a paired element, lying directly in front of the parietal, and forming with its fellow of the opposite side the anterior portion of the roof of the cranial cavity and also a considerable portion of its lateral, orbital wall. Unlike the condition in the human skull, the two bones are separate throughout life, so that there is a permanent **frontal suture.** Each consists of a **frontal portion** (pars frontalis), the external or dorsal surface of which continues that of the parietal, and of an **orbital portion** (pars orbitalis), enclosing the dorsal portion of the orbit. The two parts are connected at the **supraorbital border,** with which is also connected the base of the divided **supraorbital process.** The anterior end of the frontal portion is deeply notched where it comes in contact with the nasal and

premaxillary bones. Two processes are thus formed, one medial, the other lateral to the nasal. The medial process is associated with that of the opposite side to form a triangular **frontal spine,** while the lateral or **maxillary process** (processus maxillaris) projects forward between the nasal and premaxillary bones, on the one hand and the subcutaneous process of the lacrimal, the orbital process of the maxilla, and the body of the latter, on the other.

The **orbital portion** of the frontal forms a considerable portion of the orbital wall. Its anterior margin is in contact with the lacrimal bone, its ventral margin with the slender sphenoorbital process of the maxilla, the ethmoid process of the orbitosphenoid, and the orbitosphenoid proper. Its internal surface is divided by a vertical ridge into anterior and posterior portions, in relation respectively to the anterior and middle cranial fossae. The anterior cranial fossa is enclosed by the frontal bones, with the exception, however, of a small portion of the floor which is formed by the cribriform plate of the ethmoid.

10. THE ETHMOID BONE.

The **ethmoid bone** (os ethmoidale), the chief representative of the embryonic cartilaginous nasal capsule, is a delicate, greatly sculptured structure, almost completely enclosed by the membrane bones of the face. Its features may be studied either in the divided skull, or in one from which the roof of the nasal and cranial cavities has been removed. It consists of three main portions, namely, the **cribriform plate,** the **perpendicular plate,** and the paired **lateral masses,** or **ethmoidal labyrinths.**

The **cribriform plate** (lamina cribrosa) is exposed to the anterior cranial fossa. It is somewhat heart-shaped, with its apex in contact with the ethmoidal processes of the orbitosphenoids. Its lateral portions are perforated by numerous foramina, giving passage in the natural condition to the branches of the olfactory nerves. Its median portion forms a low vertical ridge, the **crista galli,** continuous in front with the perpendicular plate.

The **perpendicular plate** (lamina perpendicularis) is the bony, posterior portion of the nasal septum, and as such is exposed to the

nasal cavity. It is united with the cartilaginous nasal septum and also with the presphenoid. It forms the terminal member of the chain of bones lying in the basicranial axis.

The **ethmoidal labyrinth** (labyrinthus ethmoidalis) occupies for the most part the posterior portion of the nasal fossa, but the **nasoturbinal** extends forward to its anterior end, and is attached for the greater part of its length to the internal surface of the nasal bone. It is broadest in its middle portion, where it projects into the space left between the ethmoturbinal proper and the maxilloturbinal, and contains at this point a pouch-like cavity, termed the **marsupium nasale.** The whole structure is comparable to one of the folds of the ethmoturbinal proper; but it is frequently seen to be divided into anterior and posterior parts by a thin vertical line of cartilage, the anterior division being probably allied to the maxilloturbinal. Its middle, ventral, portion bears a stout, backwardly-directed **uncinate process** (processus uncinatus), which is applied to the medial surface of the maxilla.

The **ethmoturbinal proper** consists, as described above, of several shorter scrolls, decreasing in length from above downward. Like the posterior part of the nasoturbinal, they are attached directly to the cribriform plate, the perforations of which may be seen in the divided skull opening into the **ethmoidal scrolls** or spaces contained by them. They are roughly comparable to the superior and middle turbinated bones of the human skull, but in the rabbit, as in most mammals, the ethmoturbinal surfaces are relatively much more extensive than in man.

In the typical mammalian skull the ethmoid bone is exposed to the orbit, where it forms a thin plate of bone, the **lamina papyracea.** In the rabbit, however, the space usually occupied by the lamina papyracea is partly filled by the lacrimal bone, the ethmoidal process of the orbitosphenoid and the sphenoorbital process of the maxilla.

11. THE INFERIOR TURBINATED BONE.

The **inferior turbinated bone** (concha nasalis inferior), or **maxilloturbinal,** is a finely ridged structure, situated anteriorly in the nasal fossa, and supported by the maxilla and premaxilla.

It represents the similarly-named structure of the human skull, the lowermost of three scroll-like bones, of which the remaining two, the superior and middle turbinated bones, belong to the ethmoturbinal. In the natural condition it is covered by a non-olfactory epithelium, and is thus distinguishable in function as well as in position from the latter.

12. THE MAXILLA.

The **maxilla,** the largest element of the facial region, is associated with its fellow of the opposite side to form the main portion of the upper jaw. It consists of a central portion, the **body** (corpus maxillae), and of five processes, namely, **alveolar, palatine, orbital, zygomatic,** and **sphenoorbital.** In the adult condition the zygomatic bone is fused with the maxilla, so that the extent of the zygomatic process appears to be greatly increased.

The **body** of the maxilla is greatly fenestrated on its external surface, the perforated area extending backward to the anterior rim of the orbit, and thus including the **maxillary fossa** and the **infraorbital foramen.** The dorsal boundary of the bone is formed by the frontal process of the premaxilla and by the maxillary process of the frontal. Anteriorly, it is united with the premaxilla, the ventral part of the suture appearing in the diastema separating the incisors from the cheek-teeth. The ventral portion of the bone forms part of the lateral boundary of the **incisive foramen.** Behind the palatine bridge it is applied to the lateral surface of the palatine bone, and is projected into the orbit as a broad ridge enclosing the alveoli of the four posterior cheek-teeth.

In the divided skull the medial surface of the body of the maxilla is found to be concealed by the ethmoturbinal. It contains a deep longitudinal excavation, the **maxillary sinus** (sinus maxillaris), widely open to the nasal fossa, but only seen to advantage when the ethmoturbinal is removed. The lateral wall of the sinus corresponds in position to the fenestrated area of the external surface. It bears the chief part of the **nasolacrimal canal.**

The **alveolar process** (processus alveolaris) is that portion of the maxilla lodging the sockets of the cheek-teeth. In the rabbit

it is separated by the diastema, in which no teeth occur, from a corresponding but imperfectly differentiated process of the premaxilla.

The **palatine process** (processus palatinus) extends toward the median plane. It forms with its fellow of the opposite side about two-thirds of the palatine bridge.

The **orbital process** (processus orbitalis) is directed obliquely toward the dorsal surface of the skull. In conjunction with the lacrimal bone and the maxillary process of the frontal, it forms the anterior orbital rim. It is continuous with the fenestrated portion of the body, and its appearance as a process is largely due to its solid character as compared with the perforated surface lying in front of it.

The **zygomatic process** (processus zygomaticus) forms the anterior root of the zygomatic arch, and in the adult condition has fused with it the anterior end of the zygomatic bone. Its ventral angle bears a prominent **masseteric spine** for the attachment of the ligament of the masseter muscle.

The **sphenoorbital process** (processus sphenoorbitalis) lies on the medial wall of the orbit, in a position opposite to the middle portion of the ridge lodging the posterior cheek-teeth. It forms a stout buttress, the tip of which is applied to the anteroventral angle of the frontal bone. In this position it is visible from the orbit, lying between the lacrimal bone and the ethmoidal process of the orbitosphenoid.

13. THE PREMAXILLA.

The **premaxilla**, or **incisive bone** (os incisivum), forms the anterior portion of the upper jaw. It comprises a central portion, the **body**—including with the latter the scarcely differentiated **alveolar portion** containing the large and small incisors—a **frontal process**, and a **palatine process**. The body forms a portion of the palatal surface of the skull and of the lateral boundary of the incisive foramen. Its dorsal surface forms part of the boundary of the **piriform aperture**, the remaining portion of this

being formed by the nasal bone. The **palatine process** extends backward on the medial side of the bone, closely applied on the palatal surface to its fellow of the opposite side, and forms in this way a medial boundary for the incisive foramen. Its dorsal surface, in conjunction with that of the corresponding process of the other side, bears a broad **palatine groove** (sulcus palatinus), lodging a portion of the cartilage of the vomeronasal organ and nasopalatine duct. The **frontal process** (processus frontalis) is a thin bony splint, extending backward between the nasal and maxillary bones, and terminating between the former and the maxillary process of the frontal.

14. THE ZYGOMATIC BONE.

The **zygomatic bone** (os zygomaticum) is a separate element only in very young animals. In the adult it is fused anteriorly with the zygomatic process of the maxilla, the position of the original suture being roughly identifiable as the point where the free horizontal portion of the zygomatic arch arises from the transverse zygomatic process. It forms an almost sagittal plate of bone bridging the orbit and serving for the attachment of the masseter muscle of the mandible. Its dorsal margin forms posteriorly a smooth, horizontal articulation with the zygomatic process of the squamosal, the end of the bone projecting considerably behind the articulation.

15. THE NASAL BONE.

The **nasal bone** (os nasale) is a thin, elongated bone forming the roof of the nasal fossa, and, in conjunction with its fellow of the opposite side, the dorsal boundary of the piriform aperture. It is loosely articulated with the maxilla and with the bone of the opposite side by smooth (harmonic) sutures. The medial margin is supported by the dorsal edge of the nasal septum. The internal surface bears the nasoturbinal scroll.

16. THE VOMER.

The **vomer** is the median, somewhat sickle-shaped, vertical plate of bone separating the ventral portions of the nasal fossae. It is visible from the palatal surface through the incisive foramina, but its extent is best shown in the divided skull. It forms a support for the ventral border of the nasal septum, and its posterior portion bears a shelf-like projection, the **ala vomeris**, which assists in the support of the ethmoturbinal.

17. THE LACRIMAL BONE.

The **lacrimal bone** (os lacrimale) is a small element lying in the anterior wall of the orbit. It is loosely articulated with the surrounding bones, and in the dried skull is frequently missing unless care has been taken to keep it in place. It consists of a basal portion, somewhat rectangular on its orbital surface, and of two processes, namely, the **subcutaneous process** and the **hamulus lacrimalis.** The subcutaneous process is the prominent hook-like projection extending laterad beyond the orbital rim. The hamulus lacrimalis is a small process, directed toward the nasal cavity. It bears a groove which, in association with a corresponding groove of the maxillary bone, forms the first portion of the nasolacrimal canal.

18. THE PALATINE BONE.

The **palatine bone** (os palatinum) forms the poste ior portion of the palatine bridge and the major portion of the lateral wall of the nasopharynx. It consists of two portions—**horizontal**, and **perpendicular**. The **horizontal portion** (pars horizontalis) is that lying in the plane of the palatal surface. It is articulated in front with the palatine process of the maxilla, the suture between the two bones enclosing the **greater palatine foramen**, the ventral termination of the **pterygopalatine canal**. The **perpendicular portion** (pars perpendicularis) is the vertical plate extending backward from the palatine bridge. Its medial surface is divided by a low ridge into a dorsal portion, in particular relation to the nasopharynx, and a ventral portion, in relation to the oral

cavity, the ridge indicating the position of the soft palate. Its lateral surface is partly applied to the maxilla and partly exposed to the orbit. Its dorsal margin is articulated with the presphenoid and with the ethmoidal process of orbitosphenoid, but a small posterior portion is free, so that the anterior portion of the basisphenoid is visible from the orbit. The free ventral margin forms posteriorly a thick projecting angle, the **pyramidal process** (processus pyramidalis), the base of which is cleft where it articulates with the medial and lateral laminae of the pterygoid process. Between the pyramidal process and the alveolus of the last cheek-tooth there is a conspicuous **palatine notch** (incisura palatina), connecting the orbit with the palatal surface. In the entire skull only the posterior portion of the lateral surface is visible from the orbit, the anterior portion being concealed by the projecting bases of the posterior cheek-teeth. The ridge of bone on which the alveoli of these teeth are borne is separated from the palatine bone by the **infraorbital groove.** The medial wall of the latter, formed by the palatine bone, contains the orbital opening of the **pterygopalatine canal** and the **sphenopalatine foramen.**

19. THE MANDIBLE.

The **mandible** (mandibula), or lower jaw, comprises the two **dentary bones** (ossa dentalia), which, in the rabbit, as in mammals generally, are united by a fibrous or fibrocartilaginous connection (symphysis mandibulae); not coalesced, as in the human skull, to form a continuous structure. As indicated above, each of the dentary bones comprises: (1) a horizontal, tooth-bearing portion which, in conjunction with that of the opposite side, forms the **body** of the mandible (corpus mandibulae); and (2) a posterior, vertical plate, the **mandibular ramus** (ramus mandibulae), for muscle attachment and articulation. The **horizontal portion** is deep posteriorly, where it lodges the alveoli of the cheek-teeth. Anteriorly, in the diastema separating the latter from the incisors, its dorsal surface is rounded and depressed, the space thus formed corresponding to a similar space in the upper jaw and serving chiefly for the accommodation of the lips, which in this region encroach medially on the oral cavity. The medial surface of the

horizontal portion forms an acute angle with that of the bone of the opposite side, except anteriorly, where it bears a roughened area for articulation with the latter. Running backward from the symphysis there is a broad horizontal ridge, representing the **mylohyoid line** (linea mylohyoidea), the line of attachment of the mylohyoid muscle. The **mandibular foramen**, through which, in the natural condition, the inferior alveolar nerve and artery gain access to the interior of the bone, lies on this surface at the junction of the horizontal portion with the ramus. The corresponding **mental foramen** (foramen mentale), through which branches of these structures leave the mandible, is situated on the lateral surface in front of the first premolar. The mandibular foramen is closely connected with a second aperture lying at the ventral end of the sulcus ascendens, directly behind the last molar, and serving for the transmission of a vein connecting the inferior alveolar and inferior orbital veins.

The **mandibular ramus** forms in general an obtuse angle with the horizontal portion. As in other herbivores, the ventral part, distinguished as the **angle,** is greatly increased in size at the expense of the condyloid process and to a still greater extent of the coronoid process, the latter being vestigial. In addition to a low **pterygoid tuberosity** (tuberositas pterygoidea), situated at the posterior projecting point of the angle, the posterior and ventral margins of the angle are excavated on the medial side of the bone, so that they form the boundary of a pronounced, though shallow, **inferior pterygoid depression** for the insertion of the pterygoideus internus muscle. The area occupied by the pterygoideus internus is separated by a low ridge from a more dorsally placed superior depression for the pterygoideus externus muscle. A somewhat similar depression, termed the **masseteric fossa,** occupies the lateral surface of the angle, its raised ventral margin terminating posteriorly in the **masseteric tuberosity** (tuberositas masseterica). The articular portion, or **head** of the mandible is greatly elongated in the anteroposterior direction in accordance with the anteroposterior action of the lower jaw, this feature being one which is of general occurrence in the rodent order, and more fully expressed in the great extension forward and backward of the attachment areas of the muscles of mastication. The connection

of the articulating portion with the **condyloid process,** the so-called **neck** of the mandible (collum mandibulae), is a thin plate of bone, the anterior and posterior margins of which are barely notched by the **anterior** and **posterior mandibular incisures.** Connecting the anterior incisure with the rim of the alveolus of the last cheek-tooth there is a deep groove, the **sulcus ascendens,** the lateral margin of which is formed by the reduced **coronoid process** (processus coronoideus). Its low medial margin is formed by a bony stay which extends to the medial surface of the horizontal portion opposite the last cheek-tooth, and is continued forward into the mylohyoid line. The sulcus ascendens lodges in the natural condition the insertion portion of the greatly reduced temporalis muscle.

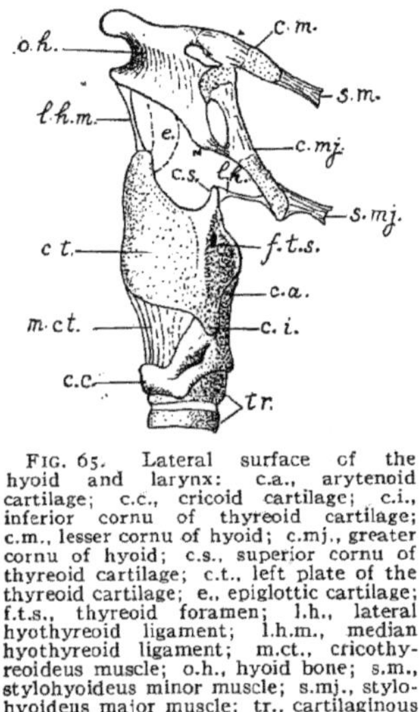

FIG. 65. Lateral surface of the hyoid and larynx: c.a., arytenoid cartilage; c.c., cricoid cartilage; c.i., inferior cornu of thyreoid cartilage; c.m., lesser cornu of hyoid; c.mj., greater cornu of hyoid; c.s., superior cornu of thyreoid cartilage; c.t., left plate of the thyreoid cartilage; e., epiglottic cartilage; f.t.s., thyreoid foramen; l.h., lateral hyothyreoid ligament; l.h.m., median hyothyreoid ligament; m.ct., cricothyreoideus muscle; o.h., hyoid bone; s.m., stylohyoideus minor muscle; s.mj., stylohyoideus major muscle; tr., cartilaginous tracheal rings.

C.

THE HYOID APPARATUS.

The **hyoid bone** (os hyoideum) (Fig. 65) is a stout, somewhat wedge-shaped bone lying in front of the larynx and between the angles of the mandible. Its ventral portion is connected with the thyreoid cartilage of the larynx by the **median hyothyreoid ligament.** With its lateral portion are articulated two independent elements, termed the lesser and greater cornua. The **lesser cornu** (cornu minus) is a small, partly cartilaginous structure, attached to the anterodorsal angle of the hyoid, and connected through the stylohyoideus minor muscle with the jugular process of the skull. The **greater cornu** (cornu majus) is a larger element extending obliquely dorsad, and similarly suspended from the jugular process by the stylohyoideus major muscle. The connection of the lesser cornu with the styloid process through the stylohyoideus minor replaces the stylohyoid ligament of the human

skull and the chain of elements commonly occurring in mammals and other vertebrates in this region. In most mammals the term "lesser", as applied to it, is inappropriate. The muscle tendon contains near the jugular process a small ossification representing a detached styloid process. This connection, together with the hyoid bone itself, indicates the relation of the embryonic hyoid arch, from which the skeletal structures in question are derived. The greater cornu belongs to the succeeding visceral arch, and is connected with the superior cornu of the thyreoid cartilage of the larynx by the **lateral hyothyreoid ligament.** It is commonly represented in mammals by a small thyreohyal process.

THE SKELETON OF THE ANTERIOR LIMB.

The skeleton of the anterior limb is divisible into two portions, namely, a proximal portion, comprising the **scapula** and the **clavicle,** and a distal portion, comprising the supports of the **free extremity.** The scapulae and clavicles of the two sides together form the **pectoral girdle.** The pectoral girdle is lightly constructed, and, apart from its muscular connections, which constitute its main support, is directly attached to the axial skeleton only through the sternoclavicular ligament.

The skeleton of the free extremity is divisible into proximal, middle, and distal segments. The proximal segment contains a single bone, the **humerus**; the middle segment two elements, the **radius** and **ulna;** while the distal segment comprises, in addition to the accessory **sesamoid bones,** twenty-eight elements of the regular series, of which nine form the **carpus,** five the **metacarpus,** and fourteen the **phalanges** of the digits.

THE SCAPULA.

The **scapula** (Fig. 66) is a somewhat triangular plate of bone lying in the natural position on the lateral surface of the anterior part of the thorax, with its apex directed downward and forward. In the rabbit, as in quadrupedal mammals generally, the main surfaces are respectively medial and lateral, and differ in this respect from the human condition, in which, from the transverse widening

of the thorax, the corresponding surfaces are more nearly ventral and dorsal. Of its three borders, one, the **superior border** (margo superior), is directed toward the occiput; another, the **vertebral border** (margo vertebralis), toward the vertebral column; and the third or **axillary border** (margo axillaris), toward the armpit. The corresponding **angles** are **medial** (superior), **inferior**, and **lateral** (glenoid). The lateral surface bears a stout bony plate, the **scapular spine** (spina scapulae), which arises from the body of the bone through about two-thirds of its extent, and ends ventrally in a free projection, the **acromion.** The posterior margin of the acromion bears a backwardly-directed process, the **meta-**

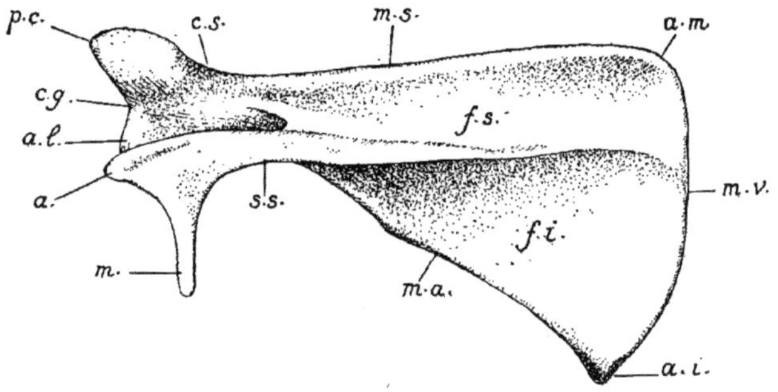

FIG. 66. Lateral surface of the left scapula: a., acromion; a.i., a.l., and a.m., inferior, lateral, and medial angles; c.g., glenoid cavity; c.s., neck of the scapula; f.s. and f.i., supraspinous and infraspinous fossae; m., metacromion; m.a., m.s., and m.v., axillary, superior, and vertebral borders; p.c., coracoid process; s.s., scapular spine.

cromion (processus hamatus). Through the presence of the scapular spine, the lateral surface of the bone is divided into two areas for muscular attachment. One of these, the **supraspinous fossa** (fossa supraspinata), lies in front of the spine, the other, the **infraspinous fossa** (fossa infraspinata), behind it. The infraspinous fossa is the more extensive one. The medial surface, on the other hand, presents a single large shallow depression, the **subscapular fossa** (fossa subscapularis), which is triangular in shape and occupies practically the entire surface. The apex or lateral angle of the scapula, sometimes termed the **head** of the bone, is expanded to a considerable extent in comparison with the

slender portion—the so-called **neck** of the scapula (collum scapulae)—connecting it with the body of the bone. It bears a concave depression, the **glenoid cavity** (cavitas glenoidalis), for articulation with the humerus. The articulating surface is borne chiefly on that part of the bone corresponding to the axillary border, but it also extends in an anterior direction to the base of an overhanging projection, the **coracoid process** (processus coracoideus). The free portion of the latter forms a blunt, hook-like projection lying toward the medial surface of the bone.

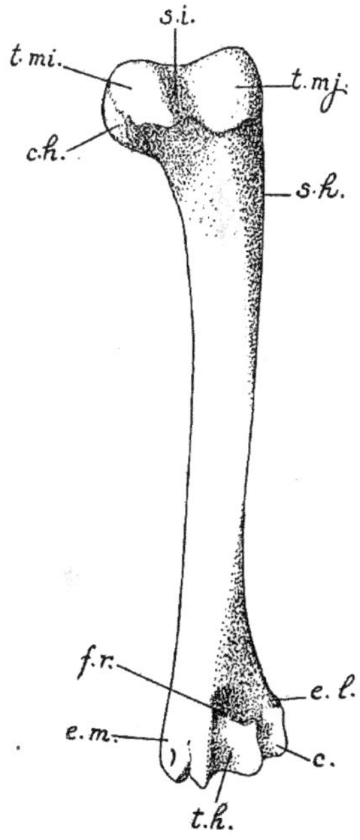

FIG. 67. Anterior surface of the left humerus: c., capitulum; c.h., head of humerus; e.l. and e.m., lateral and medial epicondyles; f.r., radial fossa; s.h., deltoid tuberosity; s.i., intertubercular groove; t.h., trochlea humeri; t.mi. and t.mj., lesser and greater tubercles.

THE CLAVICLE.

The **clavicle** (clavicula) is imperfectly developed in the rabbit, consisting of a slender, arcuate rod of bone, tipped by cartilage, which lies in the interspace between the manubrium sterni and the head of the humerus. It occupies only a portion of this interspace, being attached medially by the **sternoclavicular ligament** and laterally by the **cleidohumeral ligament**.

THE HUMERUS.

The **humerus** (Fig. 67) is typical of the long bones of the proximal and middle segments of the fore and hind limbs in consisting of a central portion, the **body** or **shaft** of the bone, and of **proximal** and **distal extremities** for muscle attachment and articulation. The proximal extremity bears on its medial side a smooth, convex projection, the **head** of the humerus (caput humeri), for articulation with the scapula. The articulation is nominally a ball-and-socket

joint, or **enarthrosis**, but the articulating surfaces are somewhat restricted, and the muscular arrangements of the limb are such that the range of lateral motion (abduction and adduction) is small. Immediately in front of the head of the bone there is a small elevation, the **medial tuberosity** or **lesser tubercle** (tuberculum minus). It is separated by a longitudinal furrow of the anterior surface, the **intertubercular groove** (sulcus intertubercularis), from a much larger lateral elevation, the **lateral tuberosity** or **greater tubercle** (tuberculum majus). Extending distad from the latter is a triangular area, the **deltoid tuberosity** (tuberositas deltoidea), the tip of which reaches almost to the middle of the bone and forms a pronounced angle on its anterior surface.

The distal extremity of the humerus bears a grooved articular surface, the **trochlea humeri**, for articulation with the radius and ulna. On its lateral side is a smaller surface, the **capitulum humeri**, for articulation with the radius alone. Immediately above the trochlea the medial and lateral portions of the bone are thickened to form two areas for muscular attachment. One of these, the **lateral epicondyle** (epicondylus lateralis), is a general point of origin for the extensor muscles of the dorsal surface of the hand, while the other, the **medial epicondyle** (epicondylus medialis), is a similar point of origin for the flexor muscles of the ventral or volar surface. Between the epicondyles the extremity of the bone is greatly excavated, so that the projecting portions of the radius in front and of the ulna behind are received into depressions of the surface when the forearm is greatly flexed or extended. On the anterior side is the **radial fossa** (fossa radialis); on the posterior side the **olecranon fossa** (fossa olecrani), so-called because it accommodates the olecranon process of the ulna.

THE RADIUS AND ULNA.

The **radius** (Fig. 68) is the shorter of the two bones of the forearm, since its proximal extremity does not extend backward beyond the front of the elbow joint. It is anterodorsal in its general position, but is crossed on the ulna in such a way that its proximal extremity tends to be lateral, while its distal extremity is medial. The proximal extremity, termed the **head** of the radius (capitulum

radii) is immovably articulated with the ulna. It bears an extensive articular surface, meeting both the trochlea and capitulum of the humerus, and thus forming a considerable portion of the elbow-joint. The body of the bone is solidly united with the ulna by the **interosseous ligament** of the forearm. The distal extremity is largely formed by an epiphysis, which is well-marked even in older animals. It bears a grooved, **carpal articular surface** (facies articularis carpea), for articulation with the navicular and lunate bones.

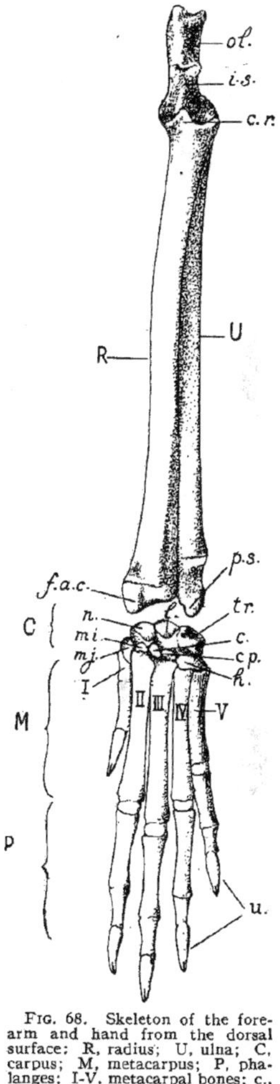

FIG. 68. Skeleton of the forearm and hand from the dorsal surface: R, radius; U, ulna; C, carpus; M, metacarpus; P, phalanges; I-V, metacarpal bones; c., central bone; cp., capitate; c.r., head of radius; f.a.c., carpal articular surface of radius; h., hamate bone; i.s., semilunar notch of the ulna; l., lunate bone; mi., lesser multangular; mj., greater multangular; n., navicular; ol., olecranon; p.s., styloid process of the ulna; tr., triquetral bone; u., ungual phalanges.

The **ulna** (Fig. 68) is a somewhat S-shaped bone, the shaft of which is vertically flattened, so that it possesses two main surfaces, respectively anterodorsal and posteroventral. The former, in conjunction with the related surface of the radius, continues the area of origin of the extensor muscles of the hand from the lateral epicondyle of the humerus distad on to the forearm, while the latter has a similar function with respect to the flexor muscles. The proximal portion of the bone is laterally compressed. It bears a crescentic depression, the **semilunar notch** (incisura semilunaris), the articulating surface of which continues that of the medial portion of the head of the radius, and is received into the trochlea humeri. Behind the elbow-joint the bone forms the large projecting portion of the elbow, the **olecranon**, which is a strong process for the insertion of the extensor muscles (anconaei)

acting on the forearm. The distal extremity of the bone is formed by an epiphysis, similar to but much longer than that of the radius. It is immovably articulated with the radius, and its tip is formed by a blunt **styloid process** (processus styloideus), which is articulated with the triquetral bone of the carpus.

The elbow-joint is formed by the trochlea and capitulum of the humerus in conjunction with the semilunar notch of the ulna and the corresponding articular surface of the head of the radius. It is a hinge-joint, or **ginglymus**, permitting motion in one plane, *i.e.*, extension and flexion of the forearm. The trochlear surface of the humerus, however, has a slight spiral trend, the anterior portion being medial in comparison with the posterior portion.

Through the immovable articulation provided by the respective proximal and distal ends of the bones, and also through the interosseous ligament, the radius and ulna are unable to change their positions with respect to one another; in other words, the radius is unable to rotate on an axis formed by the ulna, the fore foot being fixed in a position comparable to that of pronation in the human hand (cf. p. 65).

THE CARPUS.

The **carpus** (Fig. 68) comprises nine small elements, the wrist or **carpal bones** (ossa carpi), which are interposed between the forearm and the digits. They are arranged in two main rows, namely, a **proximal row**, the elements of which are articulated with the radius and ulna; and a **distal row**, the elements of which are articulated with the five bones of the metacarpus. Enumerated from the medial side of the wrist laterad, the proximal row contains four elements, namely, the **navicular, lunate, triquetral** and **pisiform bones.** The navicular and lunate are articulated with the distal extremity of the radius; the triquetral with the styloid process of the ulna. The pisiform bone lies on the ventral surface of the extremity of the ulna, and is therefore not exposed to the dorsal surface of the wrist. The distal row contains five elements, namely, the **greater multangular, lesser multangular, central, capitate,** and **hamate bones.** The first, second, and fourth are in association respectively with the first, second, and third metacarpals. The central bone lies to the

lateral side of the articulation at the base of the second metacarpal. As its name implies, it is originally an element interposed between the proximal and distal rows. The hamate is a comparatively large element associated with the fourth and fifth metacarpals, but extending also to the articulation of the third, where it tends to replace the greatly reduced capitate.

THE METACARPUS AND PHALANGES.

The **metacarpus** (Fig. 68) comprises five stout elements, the **metacarpal bones** (ossa metacarpalia), which form the basal supports of the digits. Each consists, in addition to a main portion or **body**, of a flattened proximal end, or **base**, and a rounded distal extremity, or **head**. The four lateral bones are normally developed, while the first, which belongs to a reduced digit, is of very small size.

The **phalanges** or bones of the digits are distributed according to the formula 2, 3, 3, 3, 3. They are similar in form to the **metacarpals**, with the exception, however, of the terminal, **ungual** phalanges, which are laterally compressed, pointed, and cleft at their tips for the attachment of the claws.

SESAMOID BONES.

Accessory elements, **sesamoid bones** (ossa sesamoidea) developed in connection with the ligaments of muscles, are found on the volar surface of the foot in association with certain of the joints. They occur in transverse pairs at the metacarpophalangeal articulations and in linear pairs at the articulations of the second with the third phalanges. The pisiform bone of the carpus is also a sesamoid, being formed in the insertion tendon of the flexor carpi ulnaris muscle.

THE SKELETON OF THE POSTERIOR LIMB.

In the posterior limb the proximal or girdle portion comprises the paired **coxal bones,** which are united ventrally at the pelvic symphysis, thus forming the **pelvic girdle.** The distal portion—comprising, as in the anterior limb, the supports of the **free ex-**

tremity—is divisible into proximal, middle and distal segments. The proximal segment contains a single element, the **femur;** the middle segment two elements, the **tibia** and **fibula,** which, however, are extensively coalesced; and the distal segment twenty-three elements, of which six form the **tarsus,** five the **metatarsus,** and twelve the **phalanges.**

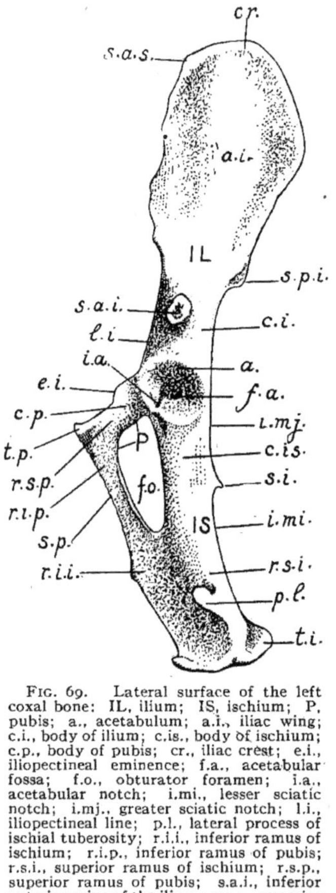

FIG. 69. Lateral surface of the left coxal bone: IL, ilium; IS, ischium; P, pubis; a., acetabulum; a.i., iliac wing; c.i., body of ilium; c.is., body of ischium; c.p., body of pubis; cr., iliac crest; e.i., iliopectineal eminence; f.a., acetabular fossa; f.o., obturator foramen; i.a., acetabular notch; i.mi., lesser sciatic notch; i.mj., greater sciatic notch; l.i., iliopectineal line; p.l., lateral process of ischial tuberosity; r.i.i., inferior ramus of ischium; r.i.p., inferior ramus of pubis; r.s.i., superior ramus of ischium; r.s.p., superior ramus of pubis; s.a.i., inferior anterior spine of the ilium; s.a.s., superior anterior spine; s.i., ischial spine; s.p., symphysis pubis; s.p.i., inferior posterior spine; t.i., ischial tuberosity; t.p., pubic tubercle.

THE COXAL BONE.

The **coxal bone** (os coxae) (Fig. 69) is a somewhat triradiate structure, the posterior limbs of which are united, so that they enclose a large aperture, the **obturator foramen** (foramen obturatum). It is firmly articulated with the sacrum, and is united with its fellow of the opposite side by a thin strip of cartilage containing a small amount of fibrous material. The latter connection is the **pubic symphysis** (symphysis pubis), better termed in the rabbit the **pelvic symphysis,** since it is somewhat more extensive than the corresponding articulation of the human pelvis.

In the young animal each half of the pelvis consists of three elements, namely, the **ilium, ischium,** and **pubis.** They form the three rays of the coxal bone, and are united with one another in the region of the

acetabulum, which is the basin-like depression for the articulation of the pelvis with the femur. Only two of the original elements, however, actually take part in the formation of the acetabulum, the pubis being excluded through the development in the acetabular depression of a small triangular element, the **os acetabuli.** Although completely coalesced in the adult condition, and showing but few traces of their original separation, the three chief elements are nevertheless described as if distinct.

The **ilium** (os ilium) is the anterior, also somewhat dorsal, portion of the bone; that part extending forward from the acetabulum. It comprises a basal portion, the **body** (corpus oss. ilium), which includes the anterior portion of the acetabulum and the cylindrical part of the bone in front of it, and an expanded portion, the **iliac wing** (ala oss. ilium), for muscle attachment and articulation with the sacrum. The body is somewhat triangular in section, its surface being divided into three areas, which are respectively medial, or **sacral,** ventrolateral, or **iliac**, and dorsolateral, or **gluteal.** The corresponding borders are respectively ventral, or **pubic**, lateral, or **acetabular,** and dorsal, or **ischial.** The acetabular border terminates a short distance in front of the acetabulum in an abruptly truncated projection, the **inferior anterior spine** (spina anterior inferior). The ischial border forms the anterior half of a long depression of the dorsal surface of the coxal bone, the **greater sciatic notch** (incisura ischiadica major). The pubic border presents on its medial side a faint, ridge-like elevation, the **iliopectineal line** (linea iliopectinea), which connects the sharp anterior border of the pubis with the articular surface for the sacrum.

The wing of the ilium forms a shovel-like expansion, the natural position of which is almost sagittal. Its lateral surface provides a fairly extensive area for the origin of the gluteal muscles. Its medial surface is a muscle surface only in its anterior portion, the posterior portion being occupied by the roughened **auricular surface** (facies auricularis), for connection with the sacrum. The dorsal margin is thin and straight. Posteriorly, where it is associated with the greater sciatic notch, there is a small projection, the **inferior posterior spine** (spina posterior inferior), also termed the tuber sacrale. Anteriorly it passes by a broad angle into the

anterodorsal margin of the bone, the latter forming the projecting end of the wing, which is distinguished as the **iliac crest** (crista iliaca). This portion is considerably thicker than the related dorsal and ventral margins, and also bears on its medial side a somewhat hook-shaped process Its anteroventral angle is the **superior anterior spine** (spina anterior superior) or tuber coxae. The ventral margin is slightly longer than the dorsal margin, and is also concave. It is associated with the pubic border of the body of the ilium, and is not connected with the inferior anterior spine.

The **ischium** (os ischii) extends backward from the acetabulum, its axis continuing that of the ilium. It consists of a basal portion, or **body** (corpus oss. ischii), a **superior ramus**, and an **inferior ramus**. The body of the ischium is for the most part cylindrical. It forms the posterior part of the acetabulum, and presents in connection with the latter a deep **acetabular notch** (incisura acetabuli), which tends to interrupt the articular surface. The acetabular notch leads forward into a depression of the centre of the articular basin, the **acetabular fossa** (fossa acetabuli). In the natural condition the combined depressions serve for the attachment of the round ligament of the head of the femur. The dorsal margin of the bone, belonging in part to the body and in part to the superior ramus, bears a short hook-like projection, the **ischial spine** (spina ischiadica). The spine divides this margin into two parts, one of which forms the posterior half of the greater sciatic notch, already described, while the other forms a similar, and, in the rabbit, scarcely less extensive, posterior depression, the **lesser sciatic notch** (incisura ischiadica minor).

The **superior** or **acetabular ramus** of the ischium is the continuation backward of the body of the bone. It is a somewhat flattened plate of bone, the thicker dorsal portion of which terminates in two blunt projections. One of these, the **ischial tuberosity** (tuber ischiadicum), forms the posterior end of the bone, while the other extends in a lateral direction and is described as the **lateral process** (processus lateralis). The **inferior** or **symphseal ramus** is that part of the ischium which extends from the superior ramus downward and forward between the obturator foramen and the symphysis to meet the corresponding ramus of the pubis.

The **pubis** (os pubis) consists of a basal portion, or **body** lying immediately below the acetabulum, a **superior** or **acetabular ramus** extending from the body to the symphysis, and an **inferior** or **symphyseal ramus** extending backward along the symphysis to its junction with the ischium. The anterior margin of the bone, described as the **pecten oss. pubis,** is thin and sharp. Near the symphysis it bears a minute elevation, the **pubic tubercle** (tuberculum pubicum), and laterally a more extensive elevation, the **iliopectineal eminence** (eminentia iliopectinea). The latter is more conspicuous in older specimens, where it is easily recognizable by its jagged outline. Its lateral margin is continuous with the iliopectineal line.

THE FEMUR.

The **femur** (Fig. 70) is a somewhat S-shaped bone, the body being very slightly arcuate, while of the two extremities, the distal one is bent downward, forming the articulation of the knee, the proximal one, with its various processes, slightly upward in association with the pelvis. In considering the general form, it will be remembered that in the natural sitting posture of the rabbit, the position of the femur is approximately horizontal, the convex surface of the shaft, which is equivalent to the anterior surface in man, being uppermost.

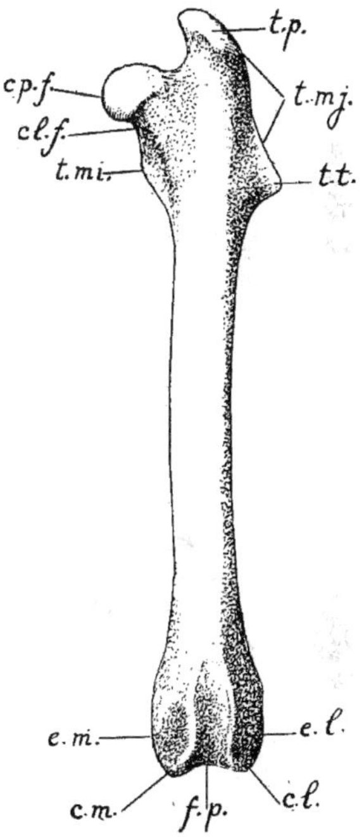

FIG. 70. Anterior surface of the left femur: c.l. and c.m., lateral and medial condyles; cl.f., neck of femur; cp.f., articular portion (head); e.l., and e.m., lateral and medial epicondyles; f.p., patellar surface; t.mi., trochanter minor; t.mj., trochanter major, including t.p. and t.t., the first and third trochanters.

The proximal extremity of the femur bears an extensive rounded portion or **head** (caput femoris), for articulation with the pelvic girdle. This portion is separated from the main part of the

extremity by a constricted area or **neck** (collum femoris), so that, unlike the case of the anterior limb, the points of muscle attachment fall a considerable distance from the point of articulation. The actual extremity of the bone is formed by a large process for muscular attachment, the **great trochanter** (trochanter major). It is divided into two portions, one of which, the **first trochanter** (trochanter primus), forms the large terminal, hook-like projection, while the other, the **third trochanter** (trochanter tertius), is the smaller lateral crest. On the medial side of the bone, immediately distal to the head, there is a triangular elevation, the **lesser,** or **second trochanter** (trochanter minor s. secundus). Posteriorly, these projections form a smooth surface for muscle attachment, except, however, at the base of the trochanter major, where the surface of the bone presents a deep, though narrow, depression, the **trochanteric fossa** (fossa trochanterica).

The distal extremity bears an extensive surface for articulation with the tibia. It is divided into two portions, known as the **medial** and **lateral condyles,** through the presence of a deep excavation, the **intercondyloid fossa** (fossa intercondyloidea). Immediately above the condyles, on the anterior surface of the bone, the intercondyloid fossa is replaced by a broad groove, the **patellar surface** (facies patellaris), which, in the natural condition, accommodates the convex internal surface of the patella. The medial and lateral portions of the bone, intervening between the distal portion of the patellar surface and the tips of the condyles, provide slightly elevated, roughened surfaces, the **medial** and **lateral epicondyles,** for muscular attachment.

THE TIBIA AND FIBULA.

The **tibia** (Fig. 71) is the larger of the two bones of the leg, lying on the medial side of the fibula, and fused with the latter for more than one-half of its length. Its proximal extremity is triangular in section, the main surfaces being respectively anterolateral, anteromedial, and posterior. The anterior border is formed by a stout ridge-like elevation, the **tuberosity** of the tibia (tuberositas tibiae), which in the natural condition serves for the insertion of the quadriceps femoris, the extensor tendon being carried over

: knee by the patella and the patellar ligament. The articular rtion is slightly differentiated into **medial** and **lateral condyles** ·responding to those of the distal end of the femur. On the :icular surface the concave areas for the reception of the condyles the femur are separated from one other by a small intervening, rtly divided hillock, the **inter-ndyloid eminence** (eminentia :ercondyloidea), and also poster-ly by a depression of the articular rder, the **posterior intercondy-ld fossa.** A corresponding **anterior tercondyloid fossa** lies in front of : intercondyloid eminence, but is orly differentiated.

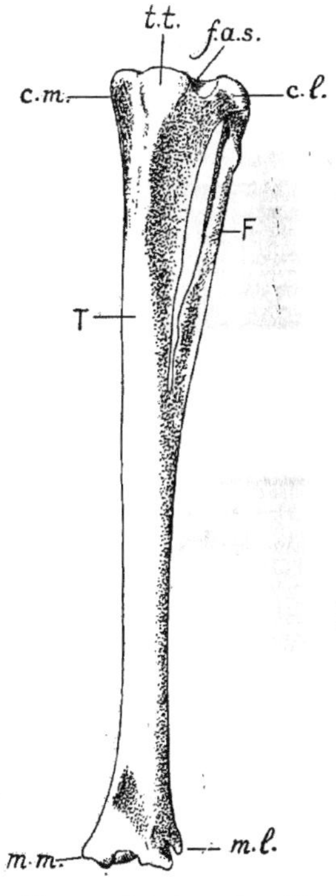

FIG. 71. Anterior surface of the left tibia (T) and fibula (F): c.l. and c.m., lateral and medial condyles; f.a.s., proximal articular surface for the femur; m.l. and m.m., lateral and medial malleoli; t.t., tuberosity of tibia.

The **fibula** (Fig. 71) is the smaller, :eral bone of the leg, and in the ɔbit is so extensively fused with the ıia that scarcely more than a third it is distinguishable. The free rtion forms a flattened bony splint, : medial margin of which is firmly ited with the tibia by the **inter-seous ligament** of the leg. Its ɔximal extremity is connected with : lateral condyle of the tibia by an ıngated epiphysis, the latter, like ɔse of the distal ends of the radius d ulna, being distinguishable even older animals.

The combined distal extremities the tibia and fibula bear a roughly :tangular articular surface for the ·sus. The tibial portion of this rface presents two grooves, separ-:d by a ridge, for articulation with the trochlea tali. On its :dial side is a small projection, the **medial malleolus** (malleolus :dialis).

The fibular portion presents a transverse groove for the convex articular surface of the calcaneus. Immediately above it, on the lateral side of the bone, is a prominent projection, the **lateral malleolus** (malleolus lateralis). It forms the anterior boundary of a groove which in the natural condition lodges the insertion tendons of the peroneal muscles.

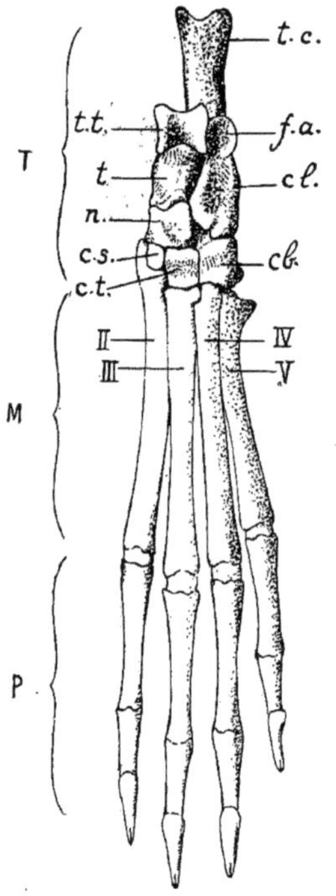

FIG. 72. The bones of the left foot, viewed from the dorsal surface: T, tarsus; M, metatarsus; P, phalanges. II-V, the four metatarsal bones: cb., cuboid; cl., calcaneus; c.s., second cuneiform; c.t., third cuneiform: f.a., articular surface for fibular side of the tibofibula; n, navicular; t, talus; t.c., tuber calcanei; t.t., trochlea tali.

THE TARSUS.

The **tarsus** (Fig. 72) comprises six elements, the **tarsal,** or **ankle-bones** (ossa tarsi), which, like the corresponding bones of the carpus, are arranged in proximal and distal rows. An exception is to be made, however, for one element, the navicular, which occupies an intermediate position. The proximal row contains two elements, the **talus** and **calcaneus**. The talus is medial and also slightly dorsal in position. Its proximal end, described as the **body** (corpus tali), bears an extensive pulley-like surface, the **trochlea tali,** for articulation with the tibia, these two surfaces together forming the chief portion of the ankle-joint. Its distal end, termed the **head** of the talus (caput tali), provides a convex articular surface for the navicular bone, and is separated from the larger trochlear portion by a slightly constricted intermediate portion or **neck** (collum tali). Its ventrolateral border is extensively articulated with the calcaneus. The latter is a cylindrical element, fully twice as long as the talus, since it is extended backward behind the ankle-joint, as the **tuber calcanei,** or bone of the heel. Its dorsal surface

bears a prominent elevation for articulation with the fibular side of the tibiofibula. Its medial surface bears a flat, shelf-like process, the **sustentaculum tali,** which forms a ventral support for the talus. The distal extremity of the bone articulates with the cuboid and also with the navicular.

The intermediate element, the **navicular bone,** is a somewhat cubical bone, lying on the medial side of the tarsus between the talus, on the one hand, and the proximal end of the second metatarsal bone and the second and third cuneiform bones, on the other. Its position is more nearly that of a central element than is the case with the bone called by this name in the rabbit's carpus. In this connection it will be remembered that the carpus and tarsus, like other parts of the limbs, are primarily constructed on the same plan.

The distal row of the tarsus contains three elements, namely, the **second** and **third cuneiform bones** and the **cuboid bone.** The two former, and especially the first, are smaller bones, articulated respectively with the second (first developed) and third metatarsals. In the rabbit the **first cuneiform bone**—the first element of the distal row in the usual condition—is fused with the proximal end of the second metatarsal. The cuboid is a larger element articulating, like the hamate bone of the carpus, with two distal elements, the fourth and fifth metatarsals. Its ventral surface bears a transverse elevation, the **tuberosity** of the cuboid (tuberositas oss. cuboidei), in front of which is a groove for the accommodation of the peculiar insertion tendon of the peronaeus primus muscle.

THE METATARSUS AND PHALANGES.

The **metatarsus** (Fig. 72) comprises five elements, of which four are fully developed and greatly exceed in size the corresponding bones of **metacarpus,** while one, the first metatarsal, is vestigial. The vestigial element lies on the plantar surface of the foot, for the most part ventral to the navicular and at the base of the second metatarsal. In each developed metatarsal there may be distinguished a main portion or **body,** a proximal extremity or **base,** and a distal extremity or **head,** the last-named portion articulating with the proximal phalanx of the digit. The base of the fifth metatarsal bears a **tuberosity** for the insertion of the peronaeus secundus muscle.

The **phalanges** are distributed according to the formula 0, 3, 3, 3, 3, the terminal, **ungual** phalanges being modified like those of the anterior limb.

SESAMOID BONES.

The sesamoid bones of the posterior limb occur at the knee-joint and on the plantar surface of the foot. On the anterior surface of the knee is the knee-pan or **patella,** through which, as indicated above, the tendon of the quadriceps femoris muscle is carried over the knee and continued as the patellar ligament to the tuberosity of the tibia. On the posterior surface there are three sesamoid bones, of which one lies in association with the medial condyle of the femur, while the remaining two are associated respectively with the lateral condyle of the femur and that of the tibia. The sesamoids of the foot are situated at the metatarso-phalangeal joints and at those connecting the second and third phalanges.

PART III.

DISSECTION OF THE RABBIT.

The plan of dissection as outlined in the following pages presupposes in the first place that the entire dissection is to be made on a single specimen, and, second, that the latter has been prepared for gross dissection according to the method given in the appendix. These points may be mentioned as explaining many details of procedure and also to a certain extent the selection in preference to others of those structures which are more readily made out by the method employed.

Because of the convenience of dissecting in circumscribed regions, the plan has been divided, although of necessity very unequally, into several parts. The order of these is such that the visceral dissection is introduced at an early stage. The somewhat more logical plan of completing first the dissection of the anterior and posterior limbs may be followed, but on account of the fact that it involves a lengthy muscular dissection to begin with, it is perhaps not to be recommended.

The account itself aims at a statement of the various structures as met with in order of dissection and the features by which they may be identified, rather than at a full description. The student should make his own observations and prove them by personal drawings and descriptions of selected parts. In this connection he will do well to bear in mind that while dissection is nominally a means of obtaining anatomical information, its chief value as a laboratory exercise consists in the training to be acquired from critical observation and analysis. It is therefore of quite as much practical importance that he should make his observations extensive and accurate as that he should employ only good instruments, or maintain the proper sequence in dissection.

I. EXTERNAL FEATURES.

The external structures, subdivisions of the body, and superficial skeletal points may be made out as follows.

1. The division of the body into the **head** (caput), **neck** (collum), **trunk** (truncus), **tail** (cauda), and the **anterior** and **posterior limbs** or **extremities** (extremitates).

2. In the head:

(a) The division into a posterior, **cranial portion** (cranium), and an anterior, **facial portion** (facies).

(b) The **mouth** (os), bounded by the cleft **upper lip** (labium superius) and the undivided **lower lip** (labium inferius). The large sensory hairs or **vibrissae.**

(c) The **nose** (nasus), and its **external apertures** (nares anteriores).

(d) The **eye** (oculus), and its coverings, the eyelids, including the **upper eyelid** (palpebra superior), the **lower eyelid** (palpebra inferior), and the **third eyelid** (palpebra tertia). The third eyelid occupies the anterior angle of the eye, and is comparable to the conjunctival fold of the human eye.

(e) The **external ear** (auricula), and its canal, the **external acoustic meatus** (meatus acusticus externus), leading to the tympanic membrane.

(f) Points on the head skeleton, to be identified by feeling through the skin; zygomatic arch, supraorbital process, external occipital protuberance, angle of the mandible, symphysis of the mandible, and the hyoid bone.

3. In the trunk:

(a) The division into **thorax, abdomen,** and **back,** or **dorsum.**

(b) The inclusion with the trunk of the proximal portions of the limbs. The angle formed by the anterior limb with the trunk represents in part the **axillary fossa** (fossa axillaris). The depression is much less evident than in man on account of the different positions of its enclosing folds formed by the pectorales and latissimus dorsi muscles. A corresponding **inguinal furrow** separates the posterior limb from the abdomen and pelvis.

(c) The **anal aperture** (anus), and on either side of it the **inguinal spaces,** hairless depressions, on which the ducts of the inguinal glands open.

(d) In the male: the **urinogenital aperture** at the extremity of the **penis**; the latter enclosed by a fold of integument, the **prepuce** (praeputium); the **scrotal sacs** (scrotum), lateral sacs of the integument lodging the testes.

(e) In the female: the urinogenital aperture, enclosed by folds of the integument, forming the **vulva.** The **clitoris,** the homologue of the penis, is contained in its ventral wall. The **mammary nipples** (papillae mammarum), eight (to ten) in number on the ventral surface of the breast and abdomen.

(f) The following skeletal points: on the axial skeleton, the manubrium sterni, xiphoid process, costal arch, spinous processes of thoracic and lumbar vertebrae; on the pectoral girdle, the acromion, clavicle, and respective borders and angles of the scapula; on the pelvic girdle, the iliac crest, pubic symphysis, and ischial tuberosity.

4. In the anterior limb:

(a) The division of the free portion into three segments, the **arm** (brachium), **forearm** (antibrachium), and **hand** (manus).

(b) The position of the **elbow** (cubitus) in comparison with the knee.

(c) The five digits, designated from the medial side as: **first** (d. primus), or **pollex, second** (d. secundus), or **index, third** or **middle** (d. tertius s. medius), **fourth** (d. quartus), and **fifth** (d. quintus s. minimus).

5. In the posterior limb:

(a) The division into three segments, the **thigh** (femur), **leg** (crus), and **foot** (pes).

(b) The **knee** (genu), and the **popliteal fossa** of its posterior surface, the latter not well defined. The projection of the heel (calx), and the angle formed by the foot with the leg.

(c) The **four** digits (dd. secundus-quintus). The vestigial first digit, or **hallux,** is not distinguishable externally.

II. THE ABDOMINAL WALL.

1. Place the animal in the supine position. Make a median incision of the skin of the ventral surface extending from the pubic symphysis to the tip of the mandible, being careful not to cut through more than the skin itself. Make three transverse incisions on the left side, the first to and along the medial surface of the arm and extending to the elbow, the second midway between the anterior and posterior limbs, the third to and along the medial surface of the thigh and extending to the knee. Work the flaps loose from the surface, using the handle of the scalpel, until the side of the trunk is well exposed. On the right side of the body it is sufficient to clear the middle line. Identify the structures as follows:

On the skin:

(a) The thick compact connective tissue forming the **corium.**

(b) The imbedded **hair-follicles.**

(c) The loose **subcutaneous tissue** (tela subcutanea) by which the skin is attached.

(d) In the female: the **mammary glands** (mammae), forming a layer on the inner surface, and more or less closely aggregated about the mammary nipples.

On the exposed surface:

(e) The **linea alba,** a white aponeurotic line extending from the pubic symphysis to the xiphoid process of the sternum.

(f) The **cutaneus maximus muscle,** a thin sheet of muscle fibres covering the entire lateral surface of the thorax and abdomen. Origin: The linea alba, the ventral surface of the sternum in its posterior portion, and the deltoid tuberosity. The last-named portion appears on the medial surface of the humerus. Insertion: The skin of the dorsolateral surface of the trunk. The fibres are directed upward and backward. The muscle is continuous across the back with its fellow of the opposite side, and is extended backward to the dorsum of the tail. It is used in shaking the skin.

The artery passing forward for a short distance in the inguinal region, and lying in the subcutaneous tissue, is the **superficial epigastric,** a branch of the femoral (p. 224). Passing into the ventral portion of the cutaneus maximus muscle, it anastomoses forward with the external

thoracic artery, a branch of the lateral thoracic. The corresponding veins are usually conspicuous in the female, since the vessels supply the mammary glands. A second anastomosis in the cutaneus muscle is formed laterally by a branch of the subscapular artery which passes backward from the axillary border of the scapula, uniting with an anterior branch of the iliolumbar artery.

The **inguinal lymph nodes** (lymphoglandulae inguinales) are small, oval, brownish bodies lying in the inguinal furrow.

2. Remove the cutaneus maximus from the surface. Identify the following points of attachment of the abdominal muscles proper:

(a) The linea alba.

(b) The **linea semilunaris**, a slightly curved line situated laterally a short distance from the linea alba.

(c) The ribs and the costal arch.

(d) The **lumbodorsal fascia** (fascia lumbodorsalis), a broad, white sheet of connective tissue extending over the posterior thoracic and lumbar regions.

(e) The **inguinal ligament** (ligamentum inguinale), a stout white cord, stretched between the symphysis pubis and the iliac crest.

3. Identify on the surface the **external oblique muscle** (m. obliquus externus abdominis). Origin: The posterior ten ribs by separate slips, the xiphoid process, and the lumbodorsal fascia. Insertion: The linea alba and the inguinal ligament. The fibres are directed from an anterior dorsal position downward and backward, the more dorsal ones almost directly backward. Some of the anterior slips of origin interdigitate with those of the thoracic portion of the serratus anterior muscle. Some are concealed by the pectoral muscles.

The muscle crossing the breast from the sternum to the arm is the pectoralis major. That passing forward from the lumbodorsal fascia to the medial surface of the humerus is the latissimus dorsi. The margins of these muscles may be raised where they conceal the external oblique.

4. Taking a line between the iliac crest and the xiphoid process, divide the muscle, and then separate it fully from the next, which may be distinguished by the cross direction of its fibres. Note the separate slips of origin and the difference in appearance between the fleshy portions of the muscle and its ventral tendinous expansion or aponeurosis; then remove it from the surface.

Examine the following muscles, proceeding in a similar manner:

(a) The **internal oblique muscle** (m. obliquus internus abdominis). Origin: The inguinal ligament, a second sheet of the lumbodorsal fascia, and the posterior four ribs. Insertion: The linea alba. The fibres pass downward and forward. The ventral aponeurosis is much broader than that of the external oblique. Near the mid-ventral line it is split into two leaves, containing between them the thin rectus abdominis muscle.

(b) The **transverse muscle** (m. transversus abdominis). Origin: Seven posterior ribs, a third sheet of the lumbodorsal fascia, and the inguinal ligament. Insertion: The linea alba. The fibres are directed downward and slightly backward.

(c) The **rectus abdominis muscle**. Origin: Lateral border of the sternum, including the xiphoid process; also the ventral surfaces of the second to seventh costal cartilages. Insertion: At the anterior end of the pubic symphysis. It is a thin, strap-like muscle, enclosed by the aponeurosis of the internal oblique, and separated from its fellow of the opposite side by the linea alba.

The artery passing forward, for the most part in this muscle, is the **inferior epigastric,** a branch of the external iliac (p. 203). It anastomoses with the internal mammary (p. 257). It gives off the **external spermatic artery,** a small vessel which perforates the abdominal wall and extends backward, supplying the sac of the testis in the male and ending in the female in the wall of the vulva.

5. Divide the remaining portion of the wall on the left side, and its whole thickness on the right, by a transverse incision, so that the viscera are fully exposed. Note on the internal surface of the wall the smooth serous investment here forming the **parietal peritoneum** (peritonaeum parietale).

III. THE STOMACH AND SPLEEN.

The cavity disclosed by the division of the abdominal wall is the **peritoneal cavity** (cavum peritonaei), the largest of the four great serous sacs representing the primary body-cavity, or **coelom** (p. 98). The major portion of the cavity is abdominal, but it extends into the pelvis, and, in the male, also into the scrotal sacs. Its lining membrane is that appearing on the body-wall as the parietal peritoneum, noted above, and on the visceral structures as the **visceral peritoneum** (peritonaeum viscerale). The visceral structures here include the major portions of the digestive and urinogenital systems.

The general relations of the visceral peritonaeum should first be examined by raising a portion of the small intestine from the left side of the visceral mass. Note its enclosure by a complete **serous coat,** similar in appearance to the membrane covering the body wall, and the extension of the latter into a **mesentery** for the support of the structure from the dorsal body-wall. Note the parallel arrangement of the arteries and veins, and also their frequent anastomoses. **Lymphatic vessels** (lacteal vessels) accompany the bloodvessels in the mesentery, but being transparent are not readily recognizable. **Lymph nodes** also occur, but in this portion of the mesentery they are aggregated at its dorsal attachment, or **root** (radix mesenterii).

For the general relations of the stomach see p. 82 and Fig. 42.

1. Displace the posteroventral portion of the liver forward, exposing in this way the ventral surface of the stomach. Without injuring the enclosing peritoneum, examine the contour of the organ and its divisions, as follows:

(a) The **greater curvature** (curvatura ventriculi major), its convex posterior surface.

(b) The **lesser curvature** (curvatura ventriculi minor), the contracted, concave anterior surface.

(c) The main portion, or **body** of the stomach (corpus ventriculi). It lies for the most part to the left of the median plane.

(d) The **cardia,** or area of junction with the oesophagus.

(e) The **fundus,** a small sac-like portion lying to the left of the cardia

(f) The **pyloric limb** (pars pylorica) forms the right portion of the organ.

(g) The **pylorus,** the point of communication of the stomach with the intestine (duodenum). It is marked by an annular constriction, preceding which is a greatly thickened muscular portion of the pyloric limb, known as the **pyloric antrum** (antrum pyloricum).

2. Raise the posterior portion of the stomach and turn it forward. Note on the dorsal surface of the greater curvature a dark red elongated body, the **spleen** (lien). On the right side of its artery, enclosed in the peritoneum, will be seen a diffuse brownish glandular mass, a portion of the pancreas. Trace the course of the peritoneum from the dorsal abdominal wall to the liver, as follows:

(a) A broad fold of peritoneum, the **mesogastrium,** connects the dorsal abdominal wall and the diaphragm with the left side and dorsal surface of the greater curvature Its posterior portion is divided into two parts by the spleen The dorsal part, the **phrenicosplenic** ligament (lig. phrenicolienale) connects the spleen with the dorsal body-wall. The ventral part, the **gastrosplenic** ligament (lig. gastroienale) connects the spleen with the greater curvature (cf. Fig. 40).

(b) The peritoneum is projected backward from the greater curvature as a free fold, the **greater omentum** (omentum majus), which covers the intestines to a certain extent. It usually contains fat. It is composed of four leaves, of which two come from the surface of the stomach and two others pass forward in a dorsal position from the free posterior border to unite with the transverse mesocolon.

(c) The **lesser omentum** (omentum minus) passes from the lesser curvature and the duodenum to the posterior surface of the liver. Its thickened margin on the right side forms the **hepatoduodenal ligament** (lig. hepatoduodenale) which carries three important structures, namely, the common bile duct, the hepatic artery and the portal vein. Its left portion forms a thin membrane, the **hepatogastric omentum,** connecting the caudate lobe with the lesser curvature.

3. Working on the left side between the dorsal surface of the stomach and the body-wall, remove sufficient of the peritoneum to expose the first portion of the abdominal aorta as it emerges from the diaphragm. Passing in the direction of the stomach is a median ventral branch, the **coeliac artery**, the distribution of which may be traced. The following structures, however, should first be identified, especially the ganglia (d, e) which are likely to be damaged in disturbing the peritoneum.

(a) The **superior mesenteric artery** (a. mesenterica superior), a second, also much larger, median branch of the aorta, given off a little distance behind the coeliac artery and passing in the direction of the intestine.

(b) The **suprarenal gland** (gl. suprarenalis) of the left side, a yellowish bean-shaped body of about half an inch in length, to the medial side of the anterior part of the kidney.

(c) The **inferior caval vein** (v. cava inferior), a large thin-walled vessel lying to the right of the aorta. It is not conspicuous if empty.

(d) The **coeliac ganglion** (g. coeliacum) of the sympathetic nervous system lies in front of the superior mesenteric artery.

(e) The **superior mesenteric ganglion** (g. mesentericum superius) of the sympathetic system lies immediately behind the superior mesenteric artery, and also on its lateral walls. The two ganglia are connected by nerves which cross the superior mesenteric artery.

(f) The nerves proceeding from the coeliac and superior mesenteric ganglia accompany the corresponding arteries, forming the **coeliac** and **superior mesenteric plexuses.**

(g) The **(greater) splanchnic nerve** (n. splanchnicus major) of the left side passes backward from its origin in the thorax (see p. 264), around the reduced left crus of the diaphragm, and crossing the aorta obliquely enters the superior mesenteric ganglion.

(h) A portion of the **pancreas** (cf. p. 189) is seen in the peritoneum after the branches of the splenic artery have been severed (4, a).

4. Trace the plan of branching of the coeliac artery, beginning at the point of origin, and exposing the vessels in order.

The **coeliac artery** (a. coeliaca) is a short trunk, its first main branch, the splenic artery, being given off near its origin from the aorta. The remaining portion of the vessel passes to the right in the direction of the lesser curvature, and divides into two parts, the left gastric and hepatic arteries. Small vessels, the **inferior phrenic arteries** (aa. phrenicae inferiores), are given off from the anterior wall of the coeliac and are distributed to the diaphragm. The distribution of the main branches is as follows:

(a) The **splenic artery** (a. lienalis) passes in the direction of the spleen, giving off small branches (**r.r. pancreatici**) to the pancreas and one or more large vessels, the **short gastric arteries** (aa. gastricae breves), to the left portion of the greater curvature. Passing along the concave surface, or **hilus**, of the spleen it gives off several **splenic branches** (rr. lienales) to that organ, and also several more branches, comparable to the short gastric arteries, to the greater curvature. Toward the end of the spleen the artery passes into the greater omentum, and at this point there is given off a large vessel, the **left gastroepiploic artery** (a. gastroepiploica sinistra), which passes to the right on the greater curvature and anastomoses with the right gastroepiploic artery.

The gastrosplenic ligament, together with its vessels, may be divided, the spleen being allowed to fall backward toward the intestine.

(b) The **left gastric artery** (a. gastrica sinistra) forms a short trunk, or more commonly a group of vessels, the branches of which pass in a somewhat radiate manner toward the lesser curvature of the stomach, reaching in this way both dorsal and ventral surfaces. Two larger vessels appear on the ventral surface respectively to the right and left of the cardia. That on the left distributes small branches (rr. oesophagei) to the oesphagus, while that on the right bears a small pyloric branch which anastomoses across the lesser curvature with the right gastric artery.

The chief nerves of the **coeliac plexus** accompany the branches of the artery to the stomach where they connect with the terminal ramifications of the vagus (p. 185).

The following artery may be identified but not traced to its termination until later, so that the bile duct and portal vein are not injured.

(c) The **hepatic artery** (a. hepatica), the continuation of the coeliac, passes forward and to the right, giving off small branches to the pancreas. Its first main branch is the **gastroduodenal artery** (a. gastroduodenalis). The latter is distributed chiefly to the first portion of the intestine as the **superior pancreaticoduodenal artery** (a. pancreaticoduodenalis superior), but a recurrent branch, the **right gastroepiploic artery** (a. gastroepiploica dextra), traverses the greater omentum to the greater curvature where it anastomoses with the left gastroepiploic artery.

After giving off the gastroduodenal artery, the hepatic enters the lesser omentum on its way to the liver. A small branch, the **right gastric artery** (a. gastrica dextra) passes to the pylorus and anastomoses across the lesser curvature with a branch of the left gastric artery.

The veins of the stomach and spleen are tributaries of the **portal vein.** Accompanying the branches of the splenic artery are the tributaries of the **splenic vein** (v. lienalis), including the **left gastroepiploic vein.** Accompanying the branches of the left gastric artery are the tributaries of the **coronary vein** (v. coronaria ventriculi). The splenic and coronary veins enter the left wall of the portal vein through a short common trunk.

On the right side of the stomach the **superior pancreaticoduodenal** vein is united with the **right gastroepiploic vein** to form a short trunk, the **gastroduodenal vein** (v. gastroduodenalis), which enters the right wall of the portal vein. The left gastroepiploic vein receives tributaries from the dorsal surface of the pyloric antrum.

The abdominal portion of the **tenth cranial,** or **vagus nerve** (n. vagus) may be traced from the oesophagus to the surface of the stomach. The left cord appears on the left wall of the oesophagus; crossing the ventral surface of the latter obliquely to the right, it ramifies on the ventral portion of the lesser curvature. The right cord passes to the stomach in a similar manner from the dorsal surface of the oesophagus.

5. Cut across the stomach at the pyloric antrum. Divide the oesophagus, and remove the stomach from the body. Open the

organ by means of an incision extending around the greater curvature to the oesophagus.

On the cut end of the pyloric antrum the **mucous** and **muscular tunics** (cf. Fig. 16) may be distinguished and separated from one another by dividing the loose tissue of the tela submucosa. On the surface of the mucous tunic may be seen the **gastric areas** (areae gastricae), formed by the longitudinal folds and imperfect transverse ridges which tend to connect them. They are well marked only in the contracted condition of the stomach. The mucous tunic of the stomach is sharply differentiated from that of the oesophagus.

IV. THE LIVER.

The **liver** (hepar) is noteworthy, first, as being the largest of the glandular structures of the body, and, secondly, as containing, in addition to the primary circulation formed by the hepatic artery and veins, the ramifications of the portal system. It is an appendage of the digestive tube, its connection with the latter being through the common bile duct.

For the general relations of the liver, see p. 23 and Fig. 42.

1. Examine the contour and plan of division as follows:

(a) The convex anterior surface, applied to the diaphragm; the concave posterior surface fitting the convexity of the stomach; the thickened dorsal portion of the organ, and its thin posteroventral margin.

(b) The division of the organ into **right** and **left lobes,** and of each lobe into **anterior** and **posterior lobules.**

(c) The **gall bladder** (vesica fellea), situated on the posterior surface of the right anterior lobule.

(d) The **quadrate lobe** (lobus quadratus), represented by one or more small divisions of the medial margins of left and right anterior lobules.

(e) The **caudate lobe** (lobus caudatus), a small independent lobe projecting backward from the base of the left posterior lobule, and accommodated in the natural condition in the space enclosed by the lesser curvature of the stomach.

(f) The **portal fissure** (porta hepatis), a large depression at the point of entrance of the portal vein.

2. Trace the peritoneal connections as follows:

(a) The lesser omentum, represented by the hepatoduodenal ligament and the hepatogastric omentum, previously divided.

(b) The **falciform ligament** (lig. falciforme hepatis), a broad median sheet connecting the anterior surface of the liver with the diaphragm and extending backward to the ventral abdominal wall. The position of this ligament indicates the line of division of the liver into right and left lobes. The free arcuate border of the ligament contains a thin cord, the **round ligament** (lig. teres hepatis), which marks the position of the umbilical vein in the foetus. The corresponding **umbilical notch** is less conspicuous than in man because of the highly lobulated condition of the whole organ in the rabbit.

(c) The **coronary ligament** (lig. coronarium hepatis), a short circular fold, continuous with the falciform, and connecting the anterior surface of the liver with the middle of the diaphragm.

(d) The **left triangular ligament** (lig. triangulare sinistrum), a lateral continuation of the coronary connecting the left lobe with the diaphragm.

3. Trace the branches of the common bile duct, the hepatic artery and the portal vein. These structures traverse the lesser omentum side by side and their branches are similarly arranged.

(a) The **common bile duct** (d. choledochus) is formed on the posterior surface of the liver by the union of a **left hepatic duct** (d. hepaticus) with a similar duct from the right anterior lobule. The latter receives the **cystic duct** (d. cysticus) from the gall bladder. Special ducts from the right posterior lobule and from the caudate lobe enter the common bile duct through a short common trunk. The common bile duct passes backward on the right side of the portal vein, and enters the digestive tube on the dorsal surface of the first (superior) portion of the duodenum immediately beyond the pylorus.

(b) The **hepatic artery** (a. hepatica) approaches the liver by passing forward on the right side of the portal vein ventral to

the bile duct. It distributes branches to the right posterior lobule and the caudate lobe, and at the base of the liver proper divides into **right** and **left rami.** The right ramus sends a branch, the **cystic artery** (a. cystica), to the gall bladder.

(c) The **portal vein** (v. portae), a vessel of large calibre, but usually found in a collapsed condition, enters the lesser omentum from the dorsal surface of the pyloric antrum. It distributes branches to the right posterior lobule and the caudate lobe; then passing directly forward to the base of the left lobe is distributed to the latter, a right branch being given off to the right anterior lobule.

4. Divide the lesser omentum with the structures described above. Divide the coronary and triangular ligaments, **cutting near the liver so as not to injure the central tendon of the diaphragm,** which resembles the coronary ligament. Remove the liver and examine its dorsal surface for the following:

(a) The **inferior vena cava**; it is accommodated in a depression of the thickened dorsal portion of the organ. The vessel should be opened lengthwise.

(b) The **hepatic veins** (vv. hepaticae) open almost directly from the substance of the liver into the inferior cava. They are typically four in number, there being separate vessels for the anterior and posterior parts of the right lobe and for the caudate lobe, in addition to a large vessel for the left lobe.

(c) The **renal impression** (impressio renalis), an extensive excavation of the right posterior lobule for the accommodation of the right kidney.

V. THE INTESTINES.

The posterior portion of the digestive tube, or that portion extending from the pyloric aperture of the stomach to the anal aperture, is divisible into two main parts, not wholly distinguishable in calibre, namely, the **small intestine** (intestinum tenue), and the **large intestine** (intestinum crassum). Both are greatly elongated and convoluted. In examining them care must be taken to avoid injury to the bloodvessels and mesenteries, especially the dorsal

attachments of the mesenteries, in which the chief plexuses and related ganglia of the sympathetic system will afterwards be traced.

(For the general relations of the intestines and mesenteries, see pp. 82, 99.)

1. Beginning at the pylorus, trace the course of the small intestine, as follows: Its first portion, the **duodenum,** forms a U-shaped loop lying on the dorsal wall of the abdominal cavity to the right of the vertebral column. The distal end of this loop, when traced from the right side, disappears in the peritoneum and may then be picked up in a forward position on the left side of the mass. This point marks the beginning of the second portion, the **mesenterial small intestine** (intestinum tenue mesenteriale), which may be traced to its termination on the greatly enlarged **caecum.** The connection with the caecum is through a rounded semi-expanded sac, the **sacculus rotundus.** The terminal portion of the small intestine is somewhat more difficult to follow on account of the adhesions of its peritoneum with that of the large intestine.

2. Examine the divisions of the duodenal loop and related structures, as follows:

(a) The **superior, descending, transverse** (horizontal), and **ascending portions** of the duodenal loop.

(b) The common bile duct, opening on the dorsal wall of the superior portion.

(c) The **mesoduodenum,** a fold of peritoneum joining the various parts of the loop.

(d) The **pancreas** (Fig. 3 B, p. 22). Its principal portion is here seen as a diffuse brownish mass lying in the mesoduodenum. Its **duct** (d. pancreatis) opens into the posterior portion of the ascending limb.

(e) The **superior pancreaticoduodenal artery,** a branch of the gastroduodenal (see p. 185), passes backward on the first portion of the descending limb.

(f) The **inferior pancreaticoduodenal artery** (a. pancreatico-duodenalis inferior), a branch of the superior mesenteric (p. 192), enters the mesoduodenum from the left side and supplies the major portion of the loop. An anterior branch anastomoses with (e).

3. In the mesenterial small intestine the following features may be identified:

(a) The lighter coloration, due to the thicker wall and greater vascularity, of the first or duodenal portion, thus distinguished as the **jejunum** (intestinum jejunum).

(b) The darker coloration, due to the thinner wall, which allows the contents to show through, and diminished vascularity of the terminal or caecal portion, thus distinguished as the **ileum** (intestinum ileum). The two portions are not distinctly separable. The **circular folds** (plicae circulares), or valvulae conniventes, of the mucous tunic, which in many mammals contribute to the thickness of the wall in the duodenum and jejunum, are, in the rabbit, not definitely expressed.

(c) The **mesentery,** the peritoneal support of the mesenterial small intestine, is distinguished in its major portion by its broad frill-like character, which allows great freedom of movement to this portion of the digestive tube. Its terminal portion, however, beginning at a point where the intestine turns sharply forward on its way to the caecum, is adherent to the mesocolon.

(d) The **mesenteric lymph glands** (lymphoglandulae mesentericae) are aggregated at the dorsal root of the mesentery, where they form a compact mass surrounding the base of the superior mesenteric artery.

(e) The wall of the sacculus rotundus shows externally a fine hexagonal pattern on account of the presence in it of a large number of lymph follicles. Similar structures, forming oval areas about 3 mm. in diameter and 5 mm. in length, may be found along the intestinal wall (**aggregated lymph nodules** of **Peyer**).

(f) The finger-like processes, or **villi,** of the mucous tunic of the small intestine may be seen by making an incision of the wall and examining its internal surface. A small portion of the wall may be excised and examined under water.

4. Trace the course of the large intestine, beginning at the sacculus rotundus, as follows:

Its first portion, the **blind intestine** or **caecum** (intestinum caecum), distinguished by its great size, is connected with the large intestine proper only in the region of the sacculus rotundus. The organ (Fig. 43) consists of three limbs, and terminates in the narrow but thick-walled **vermiform process** (processus vermiformis) or appendix. The latter lies in a dorsal position, and is directed backward.

The second portion, the **colon**, comprising the major portion of the large intestine proper, leaves the caecum in the region of the sacculus rotundus, in which position it is distinguished by its greatly sacculated walls.

The third portion, the **straight intestine**, or **rectum** (intestinum rectum) is a small terminal division situated in the middle line and enclosed for the most part by the pelvis. It is scarcely distinguishable from the related portion of the colon.

5. In the caecum the following features may be distinguished:

(a) The wall, which is otherwise smooth, is divided by a spirally arranged constriction, the latter denoting the position, internally, of a fold of the mucous tunic, the **spiral valve.**

(b) The vermiform process is a narrow, light-colored tube of about five inches in length, the wall patterned externally by lymph follicles, in the same way as that of the sacculus rotundus, and greatly thickened in comparison with that of the caecum proper.

6. The colon is divisible into ascending, transverse, and descending portions, the relations of which may be traced as follows:

(a) The **ascending colon** (colon ascendens) passes from its origin on the caecum to a point forwards on the right side of the dorsal body-wall. This portion is greatly elongated in the rabbit, and instead of passing directly forward (*i.e.*, in man, upward (see Fig. 40), follows more or less closely the course of the caecum. It is composed of five principal limbs, united by flexures. Three of the limbs are directed for the most part forward, the remaining two backward.

The first limb of the colon bears three rows of small sacculations, the **haustra**, separated by three longitudinal muscle stripes, distinguished as the **bands of the colon**

(taeniae coli). Two of these bands are free, while the third is enclosed by the supporting peritoneum, the **mesocolon.** The two free bands run together toward the anterior end of the first limb.

(b) The **transverse colon** (colon transversum) is a short segment, beginning forwards on the right and crossing the middle line transversely to the left, where it bends sharply backward, and is replaced by the descending colon.

(c) The **descending colon** (colon descendens) passes backward to a point in front of the pelvis, where it is replaced by the rectum.

The **descending mesocolon,** which connects this portion with the dorsal body-wall, should be noted on account of its relation to the inferior mesenteric artery and sympathetic plexuses. It is connected for a considerable distance with the mesentery of the ascending limb of the duodenum.

7. Displace the caecum, turning it over to the right side of the animal. Lay out the mesenterial small intestine, so that the mesentery and its bloodvessels are exposed. Remove the lymph glands from about the base of the superior mesenteric artery, first noting their position at the root of the mesentery. They receive afferent lymphatic vessels from the wall of the intestine, and send off efferent vessels to one another and to the lymphatic trunks.

Trace the branches of the superior mesenteric artery as follows:

(a) The **middle colic artery** (a. colica media), a small vessel (frequently two) arising from the left wall and passing to the transverse colon.

(b) The inferior pancreaticoduodenal artery (p. 189) arises at the same level, but from the right wall.

(c) The **ileocaecocolic artery** (a. ileocaecocolica), a large branch, equalling in size the superior mesenteric trunk, is distributed to the terminal portion of the ileum, the caecum (including the vermiform process), and the ascending colon. Its branches are arranged in two series, a **proximal group** being given off near the point of origin of the main vessel, and a **distal group,** including the terminal portion of the vessel, at about two inches from the point of origin.

The proximal branches include:

(1) Small branches to the third, fourth, and fifth limbs of the ascending colon.

(2) The **appendicular artery** (a. appendicularis) to the vermiform process. A branch of this vessel, given off near the point of origin, passes to the ileum, anastomosing with a branch of the superior mesenteric trunk.

(3) An anterior **ileocaecal artery** to the terminal fourth (third limb) of the caecum proper and related portion of the ileum.

(4) An anterior **right colic artery** to the flexure uniting the first and second limbs of the ascending colon.

(5) A posterior **right colic artery** to the second limb of the ascending colon. This vessel anastomoses with (4) and with the special branch to the third limb (1).

The distal branches include:

(6) A posterior **ileocaecal artery** to the middle portion of the third limb of the caecum and the adjacent portion of the ileum; anastomosing with (3).

(7) A **caecal artery** to the posterior portion of the third limb of the caecum.

(8) Terminal branches to the parts about the sacculus rotundus; anastomosing with (4).

(d) The **intestinal arteries** (aa. intestinales), about twenty in number, are given off from the superior mesenteric artery, and are distributed to the free portion of the mesenterial small intestine. The successive vessels are connected by anastomoses. The end of the superior mesenteric artery has the relation of one of the intestinal arteries. It anastomoses forward with a single branch given off from the side of the vessel opposite to the intestinal arteries, and the latter in turn anastomoses forwards with a branch of the appendicular artery.

8. Locate in the descending mesocolon the **inferior mesenteric artery** (a. mesenterica inferior), a small median vessel arising from the abdominal aorta. It has two main branches—the **left colic artery** (a. colica sinistra) to the anterior portion of the descending

colon (anastomosing with the middle colic), and the **superior haemorrhoidal artery** (a. haemorrhoidalis superior) to the posterior portion of the colon and the rectum.

9. The **superior mesenteric vein** (v. mesenterica superior), the chief tributary of the portal, collects the blood distributed by the superior mesenteric artery, its tributaries being similar in arrangement to the branches of the artery. The **inferior mesenteric vein** (v. mesenterica inferior) collects blood from the descending colon and rectum; it may be traced forward in the descending mesocolon.

10. Sympathetic plexuses. In the descending mesocolon will be found the **inferior mesenteric ganglion** (g. mesentericum inferius), a narrow curved body situated in front of the inferior mesenteric artery. Surrounding the abdominal aorta and appearing in the mesocolon is the **abdominal aortic plexus** (plexus aorticus abdominalis). It is connected anteriorly with the coeliac and superior mesenteric plexuses (p. 183) accompanying the corresponding vessels, and with the **renal plexuses** accompanying the renal vessels to the kidneys; posteriorly with the **inferior mesenteric** and **spermatic plexuses** about the inferior mesenteric and internal spermatic arteries, and with the **hypogastric plexus** about the pelvic vessels.

11. By dividing the rectum close in front of the pelvis, and severing the peritoneal attachments, the intestines may be separated and laid out in an extended condition. The relations to one another of the ileum, caecum, and colon are studied to much better advantage than in the natural position.

VI. THE URINOGENITAL SYSTEM.

(For the general relations of the urinogenital organs, see p. 92).

A. The Urinary Organs.

The central organs of excretion, the **kidneys** (renes), occupy an anterior position on the dorsal wall of the abdomen. The right kidney is placed a little farther forward than the left, and is largely covered by the right posterior lobule of the liver. In addition to a **fibrous coat** immediately surrounding the kidney substance, each organ is imbedded in a mass of fatty material, the **adipose capsule**

(capsula adiposa), and is also held in position by the peritoneum, which is stretched across its ventral surface.

1. By removing the peritoneum and adipose capsule from the left kidney, the external features and vascular connections may be made out as follows:

(a) The general convexity of contour.

(b) The **renal hilus** (hilus renalis), a concavity of the medial surface of the organ.

(c) The **ureter,** or duct of the kidney, a white tube passing backward from the hilus.

(d) The **renal artery** (a. renalis), arising from the abdominal aorta and entering the kidney at the hilus. A branch of this vessel, the **suprarenolumbar artery** (a. suprarenolumbalis), passes to the body-wall in front of the kidney, giving off a small **suprarenal artery** to the suprarenal gland.

(e) The **renal vein** (v. renalis), leaving the kidney at the hilus, and joining the inferior cava.

2. Divide the kidney, beginning the incision at the hilus and removing the ventral half (Fig. 50). Examine the cut surface of the dorsal half for the following:

(a) The **renal pelvis** (pelvis renalis), formed by the expanded funnel-like end of the ureter, which is fitted into the renal hilus. A central cone of kidney substance, the **renal papilla** (papilla renalis) projects into the pelvis.

(b) The **cortical substance** (substantia corticalis); distinguishable as a narrow peripheral zone of the kidney substance.

(c) The **medullary substance** (substantia medullaris), forming the central and medial portion of the kidney, including the renal papilla. It is distinguished by its radial striations.

(d) The **fibrous coat** (tunica fibrosa) of the kidney may be stripped from the surface.

In the rabbit the kidney is not lobulated. Hence there is a single renal papilla, and the division of the kidney substance into **renal pyramids** is imperfectly expressed. The medullary substance, however, possesses a slightly divided margin.

The cortical substance is of darker coloration in the natural condition, but in embalmed animals the color features are usually reversed.

3. The **urinary bladder** (vesica urinaria) lies in the ventral posterior portion of the abdominal cavity. It is a muscular sac, capable of a considerable amount of distension, but usually found in preserved animals in a greatly contracted condition. Its rounded anterior end, the **vertex,** projects forward into the abdominal cavity, while its posterior portion or **fundus,** narrows to a canal, the **urethra,** which receives on its dorsal wall the apertures of the genital ducts and those of the related glands. The connections may be made out as follows:

(a) The peritoneum is reflected from the ventral surface of the rectum in the male and from the uterus in the female, to the bladder, and after investing the latter passes to the ventral abdominal wall. The dorsal peritoneum forms in the male a double **rectovesical fold** (plica rectovesicalis), and in the female a similar **vesicouterine fold,** a recess of considerable extent (**rectovesical** or **vesicouterine pouch**) being left between the adjacent structures.

The ventral peritoneum forms a broad median vertical sheet, the **middle umbilical fold** (plica umbilicalis media). The free edge of this fold, extending from the vertex of the bladder to the umbilicus, contains a slender cord, the **middle umbilical ligament** (lig. umbilicale medium). The latter marks the position of the peripheral portions of the umbilical arteries in the foetus.

(b) The **umbilical artery** (a. umbilicalis), a branch of the hypogastric, passes along the side of the bladder to the vertex. From the base of the artery branches are given off to the ureter (a. ureterica) and related portions of the genital ducts.

B. The Male Genital Organs.

1. Continue the median ventral incision of the skin backward along the symphysis to the penis. Reflect the skin on both sides to clear the attachments of the penis to the ischium, and on the left to a point beyond the scrotum. Note the **cremaster muscle** (m. cremaster), a thin layer of muscle fibres forming the outer layer of the sac of the testis. It is continuous with the internal oblique muscle of the abdominal wall, and also contains fibres

from the transverse muscle. It is supplied with blood from the external spermatic artery. Make a longitudinal incision through this muscle, cutting forward into the abdominal cavity. On spreading apart the two flaps the following features may be made out:

(a) The **parietal layer** (lamina parietalis) of the **tunica vaginalis propria,** a layer of peritoneum, continuous with that of the abdominal wall, forms the internal lining of the sac of the testis (cf. p. 100 and Fig. 55). The sac is widely open to the abdominal cavity so that the testis passes freely from one cavity to the other.

(b) The male reproductive gland, the **testis**, with its associated vessels and duct, occupy the cavity of the sac, the testis being suspended from its dorsal wall.

(c) The **gubernaculum,** a short thick cord containing smooth muscle fibres, joins the posterior end of the testis with the end of the sac.

(d) The **visceral layer** (lamina visceralis) of the **tunica vaginalis propria** forms the peritoneal coat of the testis and is continuous with the **mesorchium,** a broad vertical fold of peritoneum connecting the testis dorsally and anteriorly with that of the body-wall.

(e) The first portion of the duct of the testis, the **epididymis,** usually imbedded in fat, forms a thickened mass at the anterior end of the testis, and extends backward as a thinner cord along its side. The thickened anterior portion is the **caput epididymidis** and the contracted posterior portion, the **cauda epididymidis.** The second portion, the **ductus deferens,** leaves the testis at its posterior end, where it is firmly attached to the gubernaculum. The connection with the epididymis may be shown by carefully separating the duct from the gubernaculum and the side of the testis. The ductus deferens receives its blood supply from the umbilical artery.

(f) The **internal spermatic artery** (a. spermatica interna) arises from the abdominal aorta, in the neighbourhood of the inferior mesenteric artery, or opposite the sixth lumbar vertebra, the left artery usually behind the right. It enters the anterior end of the testis.

(g) The **spermatic vein** (v. spermatica) is formed by a network of vessels, the **plexus pampiniformis,** which surrounds the internal spermatic artery in its posterior portion. It opens forwards into the inferior caval at the angle formed by the latter with the renal artery. (That of the right side enters the inferior caval at about the level of the spermatic arteries.)

Owing to the open communication of the testis sac with the abdominal cavity, the association of the ductus deferens with the spermatic vessels to form a **spermatic cord** (funiculus spermaticus) is very imperfectly expressed.

2. The connections of the deferent ducts with the common urinogenital tube and related parts may be displayed by dividing the symphysis and dissecting in the interior of the pelvis. The structure and attachments of the penis should first be examined.

Apart from its terminal portion, the **glans penis,** and the urethra, the body of the penis is chiefly formed by a pair of hollow fibrous structures, the **cavernous bodies** (corpora cavernosa penis). Each of these is firmly attached to the posterior margin of the ischium by a white fibrous cord, the **crus penis.** The latter is partly concealed by a short thick **ischiocavernosus** muscle, the origin of which occupies a similar position on the ischium. The penis is also attached to the symphysis by a short but stout **suspensory ligament** (lig. suspensorium) and by a thick spindle-shaped **pubocavernosus muscle** lying between the ischiocavernosi.

The attachments of the penis should be severed at the posterior border of the ischium and the symphysis divided. By pressing apart the two sides of the pelvis the urinogenital tubes, with the penis and the pelvic portion of the rectum, may be dissected out and removed from the body.

The **middle haemorrhoidal artery** (a. haemorrhoidalis media), a branch of the hypogastric, passes to the side of the rectum, to the urethra, and to the seminal vesicle. The **internal pudendal artery** (a. pudenda interna), accompanied by the corresponding nerve and vein, passes to the side of the penis, giving off the **inferior haemorrhoidal artery** to the terminal portion of the rectum and to the associated **rectal** or **anal gland.** The rectum is connected with the root of the tail by the **rectocaudalis muscle,** a somewhat spindle-shaped aggregation of smooth muscle fibres, arising from the body of the second caudal vertebra, and inserted a short distance forwards on the dorsal surface of the

rectum. The **sphincter ani externus** and **sphincter ani internus** are two closely related muscles enclosing the rectum and urethra, the former arising from the dorsum of the tail.

The following parts of the urinogenital system may be made out:

(a) The connection of the bladder with the outside of the body through the urethra. It comprises a short **prostatic portion** in relation to the genital ducts, a much longer **membranous portion** traversing the pelvis, and a terminal **cavernous portion** in the penis.

(b) The **seminal vesicle** (vesicula seminalis) lies on the dorsal surface of the base of the bladder. It is a flattened median pouch, the anterior, slightly divided tip of which is directed forward.

(c) The deferent ducts lie between the seminal vesicle and the dorsal wall of the bladder. They terminate in the ventral wall of the seminal vesicle.

(d) The **prostate** (prostata), a white or yellowish mass of glandular tissue, lies in the posterior portion of the dorsal wall of the seminal vesicle. Its ducts, three or four in number on either side, open into the urethra. Accessory **paraprostatic glands** (gl. paraprostaticae), minute finger-like projections of the urethral wall, lie on either side of the base of the seminal vesicle.

(e) The **bulbourethral gland** (gl. bulbourethralis); situated on the dorsal wall of the urethra behind the prostate.

3. The internal surface of the dorsal wall of the urethra may be exposed by a longitudinal incision extending into the bladder. The crescentic aperture of the seminal vesicle lies immediately in front of an oval elevation, the **colliculus seminalis**, on either side of which some of the minute apertures of the prostate may sometimes be made out.

C. The Female Genital Organs.

1. The organs may be traced from the abdominal cavity backward, as follows:

(a) The **ovary** (ovarium) is a small—in young animals minute—elongated structure of greyish or yellowish coloration lying

on the dorsal body-wall some distance behind the kidney. It is readily distinguished by the circular translucent dots representing the larger **vesicular ovarian follicles.** In some cases the darker radiate impressions (**corpora lutea**) left by extruded eggs are discernible.

(b) The **mesovarium,** a short fold of peritoneum suspending the ovary from the body-wall.

(c) The **internal spermatic artery** (a. spermatica interna) arises from the abdominal aorta, immediately behind the origin of the inferior mesenteric artery, and crosses the body-wall transversely to the ovary.

(d) The **spermatic vein** (v. spermatica) leaves the medial side of of the ovary, and crossing the body-wall enters the inferior caval vein.

(e) The **uterine tube** (tuba uterina), the first portion of the oviduct, distinguishable by its narrow calibre, opens into the abdominal cavity through a broad funnel-like expansion, the **ostium abdominale tubae uterinae.** The margin bears a large number of short folds and processes, the **fimbriae tubae,** which tend to enclose the margin of the ovary. A single cyst-like **hydatid** may be seen in the funnel-like expansion of the tube, but in embalmed animals is usually found collapsed.

(f) The **mesosalpinx** is the peritoneum supporting the uterine tube. It is continuous with the mesovarium.

(g) The **uterus,** the second portion of the oviduct; distinguished by its greater diameter and muscular walls. The size of this portion is enormously increased in animals which contain or have borne young.

(h) The **mesometrium** is the supporting peritoneum of the uterus, and is a continuation of the mesosalpinx. The mesometrium, mesosalpinx and mesovarium together constitute the **broad ligament** (lig. latum uteri).

(i) The **ovarian ligament** (lig. ovarii proprium) crosses the mesosalpinx, connecting the ovary with the anterolateral end of the uterus. It is continued by:

(j) The **round ligament** (lig. teres uteri) which may be traced from the anterior end of the uterus to the body-wall below the posterior portion of the inguinal ligament, where it is inserted into a small peritoneal recess, the homologue of the testis sac of the male.

(k) The **vagina** is a flattened median tube with muscular walls; it receives anteriorly the apertures of the right and left uteri.

2. Preparatory to dissecting the urinogenital structures of the pelvis, the median incision of skin of the ventral surface should be continued backward to the base of the clitoris, the structure and attachments of the latter being then examined as follows:

(a) The **cavernous bodies** (corpora cavernosa clitoridis) form the body of the structure, as in the male, and are attached to the posterior border of the ischium through short fibrous cords, the **crura clitoridis.**

(b) The **suspensory ligament** (lig. suspensorium clitoridis) is a short median cord joining the base of the clitoris with the posterior end of the symphysis.

(c) The **pubocavernosus** and **ischiocavernosus muscles** pass from the posterior border of the ischium to the base of the clitoris on either side, the former being median in position.

The attachments of the clitoris should be severed and the symphysis divided. By pressing apart the two sides of the pelvis the urinogenital tube may be dissected out and removed, together with the terminal portion of the rectum.

The related structures, the middle haemorrhoidal artery, etc., appearing in this dissection are as in the male (see note p. 198).

In the urinogenital ducts, examine the extent of the vagina backward and its connection with the canal of the bladder to form the common **vestibulum.** The latter is comparable to the male urethra (cf. p. 91, Fig. 48). The **bulbourethral gland** (gl. bulbourethralis) (cf. p. 199) lies on the dorsal wall.

By slitting open the vestibulum and extending the incision into the bladder and also forward into the left uterus the apertures of these structures may be examined from the interior. There is an **external uterine aperture** (orificium externum uteri) for each division of the uterus.

VII. THE ABDOMINAL AORTA, INFERIOR CAVAL VEIN, AND SYMPATHETIC TRUNKS.

The dissection and removal of the intestines and urinogenital organs clears the dorsal body-wall for an examination of the abdominal portion of the aorta, the inferior caval vein and the sympathetic trunks. If the inferior cava does not contain blood its tributaries should be cleared first, in order to keep them from being damaged; otherwise the branches of the aorta should first be traced. The anterior portion of the inferior cava has been removed with the liver.

1. The abdominal portion of the aorta, described as the **abdominal aorta** (aorta abdominalis) extends from the hiatus aorticus of the diaphragm to the seventh lumbar vertebra, where it is replaced by the paired common iliac arteries. It passes backward in a median position along the ventral surfaces of the bodies of the vertebrae. Its natural continuation backwards on the sacrum and the caudal vertebrae is represented by the greatly reduced median sacral artery.

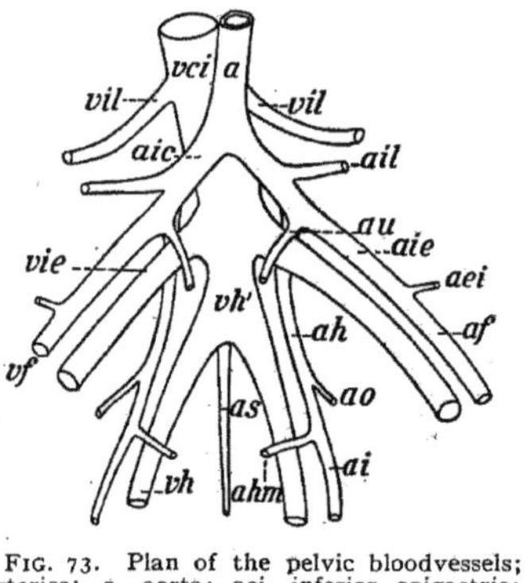

FIG. 73. Plan of the pelvic bloodvessels; Arteries: a, aorta; aei, inferior epigastric; af, femoral; ah, hypogastric; ahm, middle haemorrhoidal; ai, sciatic; aic, common iliac; aie, external iliac; ail, iliolumbar; ao, obturator; as, sacral; au, umbilical. Veins: vci, inferior cava; vf, femoral; vh, hypogastric; vh', common hypogastric; vie, external iliac; vil; iliolumbar.

The branches of the vessel are distributed in two series: (1) **visceral branches** (rami viscerales) to the parts of the digestive tube and the urinogenital organs; and (2) **parietal branches** (rami parietales) to the body-wall.

The visceral branches comprise the paired **renal** and **spermatic** arteries, and the unpaired **coeliac, superior mesenteric** and **inferior mesenteric** arteries, which have already been traced.

The parietal branches comprise:

(a) The **superior phrenic arteries** (aa. phrenicae superiores), small vessels arising by a common trunk in the hiatus

aorticus and passing to the diaphragm (usually seen better on the right side).

(b) The **suprarenolumbar artery** (a. suprarenolumbalis), arising on either side from the renal artery, and passing antero-laterad to the body-wall, supplying also the suprarenal body.

(c) The **lumbar arteries** (aa. lumbales), seven pairs of vessels distributed metamerically to the lumbar portion of the body-wall. Six pairs arise from the dorsal wall of the aorta, the seventh from the median sacral artery. They arise as single trunks, branching into right and left.

(d) The **median sacral artery** (a. sacralis media) arises from the dorsal wall of the aorta near its posterior end, and passes backward on the ventral surface of the sacrum in the middle line. Its first portion is concealed from the ventral surface by the common hypogastric vein.

2. The **common iliac artery** (a. iliaca communis) is a short trunk, the branches of which pass to the posterior limb, the wall of the pelvis, and the pelvic viscera. After giving off the **iliolumbar artery** (a. iliolumbalis), which passes laterad to the body-wall, the common iliac divides into two branches the connections of which may be traced as follows:

(a) The **external iliac artery** (a. iliaca externa) is the larger lateral branch, directed toward the inguinal ligament, over which it passes to the medial surface of the limb, becoming the femoral artery. Near its crossing with the ligament it gives off the **inferior epigastric artery** (a. epigastrica inferior), the main portion of which passes forward in the medial portion of the abdominal wall.

(b) The **hypogastric artery** (a. hypogastrica) is the smaller medial branch, directed backward on the dorsal wall of the pelvis. Its course may be traced without injury to the nerves of the lumbosacral plexus. At its junction with the external iliac the vessel gives off the **umbilical artery** (a. umbilicalis) to the bladder, or in the female first to the vagina and uterus (a. uterina). The **obturator artery** (a. obturatoria) passes laterad to the pelvic wall and the **middle haemorrhoidal** to the side of the rectum and

urethra. The main vessel leaves the pelvic cavity as the **sciatic artery** (a. ischiadica), passing to the lateral side of the abductor caudae anterior, and reappears posteriorly, dividing into the **internal pudendal** and **lateral caudal** arteries.

3 The **inferior caval vein** (v. cava inferior) is formed on the dorsal surface of the posterior end of the aorta by the union of the paired external iliac veins with the common hypogastric. From this position it passes to the right side of the aorta (rarely to the left) almost to its ventral surface, and then runs forward on the right side to the diaphragm. Its **visceral roots** or **tributaries** (radices viscerales) comprise the paired **renal** and **spermatic veins**, and the **hepatic veins** from the liver (p. 188). Its **parietal tributaries** (radices parietales) include the **inferior phrenic veins** (vv. phrenicae inferiores), which enter the inferior cava from either side of the diaphragm, the **lumbar veins** (vv. lumbales), a series of vessels corresponding to the lumbar arteries, and the paired **iliolumbar vein** (v. iliolumbalis).

4. The **external iliac vein** (v. iliaca externa), the continuation of the femoral vein of the thigh, approaches the inferior cava from the dorsal side of the inguinal ligament. It receives the **inferior epigastric vein** from the abdominal wall and the **vesical vein** from the bladder, the latter including in the female also the veins of the uterus.

5. The **sympathetic trunk** (truncus sympathicus). Its lumbar and sacral portions, and, with due care, its caudal portions may be traced on either side by working between the abdominal aorta (or its continuation, the median sacral artery) and the body-wall. Except on the ventral surface of the sacrum, the ganglia of opposite sides lie close together. The **lumbar portion** of each trunk comprises seven ganglia with their connections. The ganglia lie on the lateral surfaces of the lumbar arteries near the points where the latter disappear dorsally in the body-wall. The **rami communicantes** may be found passing from the ganglia toward the spinal nerve-roots. The **sacral portion** comprises four ganglia of which the first is much larger than the others. The **caudal portion** comprises two minute ganglia and an unpaired terminal ganglion uniting the two trunks.

VIII. THE ANTERIOR LIMB.

For this dissection the skin must first be reflected from the lateral surface of the limb and the side of the neck to the dorsal median line. It is advisable to divide the skin at the elbow, leaving the forearm and hand covered, so that the tendons of the muscles do not become dried out before they can be examined.

Covering the side and ventral surface of the neck is a broad thin sheet of muscle, the **platysma**, replacing the cutaneus maximus of the trunk. It forms a continuous layer over the dorsal surface of the neck, at which place it is also continuous with the cutaneus maximus. Passing forward from the manubrium sterni is a narrow band of fibres, inseparable from the platysma but lying beneath it, the **depressor conchae** (parotideoauricularis) **posterior**, which is inserted into the external base of the ear. The entire sheet of muscle is so closely attached to the skin that it is sometimes removed with the latter. If in place it should be raised from the surface, separated posteriorly from its attachment, and turned forward on the head.

The dissection is mainly muscular, but the arteries and nerves should be kept intact for later examination.

Identify the manubrium sterni. The muscle directed forward from it toward the angle of the mandible is the sternomastoideus, one of the muscles of the head. The external jugular vein lies on its lateral side. Identify the clavicle rudiment and the cleidohumeral ligament; then proceed to uncover the muscles directed forward from this region to the head.

For the general muscle relations of the limbs, see p. 64.

*1. Muscles arising from the axial skeleton and inserted on the scapula and clavicle.

(a) The **cleidomastoideus.** Origin: Mastoid portion of the skull. Insertion: Middle portion of the clavicle.

The muscle lying on its medial side and arising from the manubrium sterni is the **sternomastoideus,** one of the muscles of the head.

*The structures of Group 2 may be dissected first if preferred, the serratus anterior muscle being exposed from the lateral surface and divided together with the latissimus dorsi.

(b) The **basioclavicularis** (basiohumeralis). Origin: Basioccipital bone. Insertion: Lateral third of the clavicle and the cleidohumeral ligament.

(c) The **levator scapulae major**. Origin: Cartilage union of basioccipital and basisphenoid (sphenooccipital synchondrosis). Insertion: Metacromion.

The **superficial cervical artery** (p. 257) passes obliquely forward and outward under cover of these muscles, ramifying beneath the superior portion of the trapezius in the fat-mass of the side of the neck. Its **ascending cervical** branch passes forward on the lateral surface of the external jugular vein.

(d) The **trapezius**. Origin in two portions:

Superior (cervical) **portion**: External occipital protuberance and dorsal ligament of neck (ligamentum nuchae). Insertion: Metacromion and supraspinous fascia.

Inferior (thoracic) **portion.** Origin: Spinous processes of the thoracic vertebrae and the lumbodorsal fascia. Insertion: Dorsal half of the scapular spine. The muscle forms a broad triangular sheet on the dorsolateral surface of the shoulder.

The levator scapulae major, basioclavicularis, and trapezius should be divided. On the ventrolateral surface of the superior portion of the trapezius and levator scapulae major may be found nerves from the ventral rami of the third, fourth and fifth cervical spinal nerves. The **great auricular nerve** (n. auricularis magnus) passes from the third to the ear.

(e) The **rhomboideus minor**. Origin: Ligamentum nuchae. Insertion: Anterior two-thirds of the vertebral border of the scapula.

(f) The **levator scapulae minor.** Origin: Mastoid and supraoccipital portions of the skull. Insertion: Medial surface of the inferior angle of the scapula.

(g) The **rhomboideus major** Origin: Spinous processes of the first seven thoracic vertebrae. Insertion: Posterior third of vertebral border. The rhomboidei are almost continuous.

By dividing the rhomboidei the scapula may be displaced laterad. The operation is facilitated by dividing the latissimus dorsi, the relations of which should, however, first be noted (2, a).

(h) The **serratus anterior.** Origin in two portions:
Cervical portion: Transverse processes of the posterior five cervical vertebrae and the anterior two ribs.
Thoracic portion: Third to ninth ribs by separate slips alternating with those of the external oblique.
Common insertion: Medial surface of the vertebral border of the scapula.

The transverse artery of the neck (a. transversa colli) lies on the medial side of the cervical portion.

2. Muscles arising from the axial skeleton and the pectoral girdle and inserted on the humerus, for the most part at its proximal extremity.

Note the **axillary lymph glands** lying in the fat of the axillary fossa.

(a) The **latissimus dorsi.** Origin: Lumbodorsal fascia and four posterior ribs. Insertion: Deltoid tuberosity. A long flat triangular muscle, covering a considerable portion of the lateral surface of the thorax; having its dorsal angle covered by the thoracic portion of the trapezius. Its insertion end passes to the medial side of the humerus.

(b) The **pectoralis primus** (p. tenuis). Origin: Manubrium sterni. Insertion: Deltoid tuberosity.

A branch of the thoracoacromial artery appears between this muscle and the deltoideus.

The muscle should be raised from the surface and divided.

(c) The **pectoralis secundus** (p. major). Origin: Entire lateral portion of the sternum. Insertion: Anteromedial surface of the humerus, beginning below the head, and extending to the boundary between the middle and distal thirds. The anterior fibres are covered by those of (b). Some of the posterior fibres are inserted highest on the humerus.

By working backward from the clavicle the muscle can be detached from those underneath and divided.

(d) The **pectoralis tertius** (p. minor). Origin consisting of two portions:
First portion: Anterior portion of sternum to attachment of fourth rib.
Second portion: Manubrium sterni and costal cartilage of first rib.
Insertion: The superficial fibres of the first portion are attached to the clavicle. The remaining fibres, combined with those of the second portion and those of the pecto-scapularis, pass to the dorsal side of the clavicle and over the shoulder to be inserted on the scapular spine. The muscle forms a broad fleshy mass covering the anterodorsal portion of the shoulder.

(e) The **pectoralis quartus.** Origin: The sternum, from the attachment of the fourth to seventh costal cartilages. Insertion: Anterior surface of the head of the humerus, passing thence to its medial side. The muscle overlaps the more posterior fibres of the first portion of (d).

(f) The **pectoscapularis.** Origin: The manubrium sterni at the point of attachment of the first costal cartilage. Insertion as indicated above. A slender muscle concealed by the p. tertius, which should be divided.

BLOODVESSELS AND NERVES OF THE AXILLARY FOSSA.

After division of the pectorals and the clavicle, the bloodvessels and nerves of the axillary fossa will be fully exposed.

*The **axillary artery** (a. axillaris), the continuation of the subclavian, crosses from the first rib to the medial surface of the humerus, where it is replaced by the brachial artery. Its branches are:

(a) The **transverse scapular** (suprascapular) **artery** (a. transversa scapulae); it arises from the anterior wall, and taking a position dorsal to the clavicle accompanies the p. tertius and pectoscapularis to the front of the shoulder, where, under cover of these muscles, it passes into the supraspinatus muscle.

(b) The **thoracoacromial artery** (a. thoracoacromialis); it arises from the ventral wall or in common with (c), passes between the pectorales tertius and quartus, then between the p. superficialis and the cleidohumeralis. It distributes branches to these muscles, and taking a position ventral to the clavicle, passes to the platysma and the skin.

(c) The **lateral** (long) **thoracic artery** (a. thoracalis lateralis) arises from the posterior wall or in common with (b); distributes branches chiefly to the p. secundus, and sends a long superficial branch, the **external thoracic artery,** backward through the cutaneus maximus muscle. This vessel is usually conspicuous in the female, where it distributes **external mammary branches** to the mammary glands. It anastomoses posteriorly with the superficial epigastric branch of the femoral.

(d) The **subscapular artery** (a. subscapularis) is a large branch given off from the distal portion of the axillary artery. It distributes branches to the subscapularis muscle, and sends a **thoracodorsal branch** into the latissimus dorsi. Perforating the teres major muscle near the axilla, it appears on the lateral surface of the shoulder, where it sends a large branch into the inferior portion of the trapezius, and a second into the cutaneus maximus. The latter vessel supplies the proximal end of the long head of the triceps, but its chief portion passes backward uniting with an anterior superficial branch of the iliolumbar, and thus forming one of three superficial anastomoses covering the abdominal region and in the female the mammary glands.

(e) The **circumflex arteries** of the head of the humerus, see p. 215.

(f) The **deep artery** (a. profunda brachii), see p. 215.

The **axillary vein (v. axillaris) begins at the medial side of the humerus and crosses the axillary fossa to the first rib whence it is continued as the subclavian. It receives the **lateral thoracic** and **subscapular veins,** which accompany the corresponding arteries, and also the **cephalic vein** (p. 216), which reaches the medial side of the shoulder from the anterior surface of the arm by passing between the teres major and subscapular muscles near the neck of the scapula.

***The **brachial plexus** (plexus brachialis) is the network of nerves formed from the ventral rami of the posterior five cervical and first thoracic spinal nerves. The cervical nerves also take part in the formation of the more general **cervical plexus** embracing all nerves of the cervical series. The strands of the brachial plexus cross the axillary fossa and at the medial surface of the humerus are largely replaced by the three chief trunks of the free extremity, the **radial, median,** and **ulnar** nerves. These nerves are formed principally from the eighth cervical, but adjacent nerves also contribute, especially a large bundle which enters the radial from the seventh cervical, from which combined trunk the stout **subscapular nerves** are given off to the corresponding muscle. A **suprascapular nerve,** formed chiefly from the sixth cervical, passes to the anterior border of the scapula, entering the supraspinatus muscle.

By dividing the axillary nerves and vessels and the two parts of the serratus anterior muscle, the limb may be removed from the body.

4. Muscles arising from the pectoral girdle and inserted on the humerus. These muscles act on the humerus through the shoulder-joint, and except for the unimportant difference in origin are similar to those of Group 2.

The course of the cephalic vein (p. 216) should be traced before separating the muscles of the front of the forearm.

Note the **supraspinous** and **infraspinous fasciae** covering the corresponding portions of the lateral surface of the shoulder.

(a) The **cleidohumeralis.** Origin: Lateral portion of the clavicle and the cleidohumeral ligament. Insertion: Anterior surface of the humerus in its distal third. The muscle is a continuation of the basioclavicularis, but represents the brachial part of the **brachiocephalic** muscle, which, in many mammals with reduced clavicle, extends from the mastoid portion of the skull to the front of the arm. The cervical part is the cleidomastoideus.

(b) The **deltoideus**:

Acromial portion. Origin: The Acromion. Insertion: Distal portion of deltoid tuberosity.

Scapular portion. Origin: Infraspinous fascia. The end of the muscle forms a curved line over the dorsal portion of the infraspinatus, leaving only a small triangular portion of the latter exposed. Insertion: The distal portion of the muscle passes beneath the metacromion, which also serves as a point of attachment, and is replaced on the lateral surface of the humerus, beneath the acromial portion, by a long thin tendon, through which it is inserted.

The scapular portion of the deltoideus should be separated from the infraspinatus and divided, the distal end being reflected together with the metacromion.

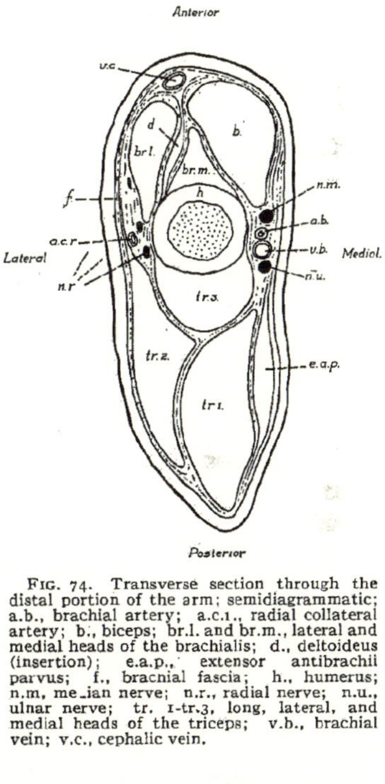

FIG. 74. Transverse section through the distal portion of the arm; semidiagrammatic; a.b., brachial artery; a.c.r., radial collateral artery; b., biceps; br.l. and br.m., lateral and medial heads of the brachialis; d., deltoideus (insertion); e.a.p., extensor antibrachii parvus; f., bracnial fascia; h., humerus; n.m., median nerve; n.r., radial nerve; n.u., ulnar nerve; tr. 1-tr.3, long, lateral, and medial heads of the triceps; v.b., brachial vein; v.c., cephalic vein.

(d) The **infraspinatus**. Origin: Posterior portion of the lateral surface of the scapula, including the spine. Insertion: Greater tubercle of the humerus. The muscle fills the infraspinous fossa.

(e) The **supraspinatus.** Origin: Anterior portion of the lateral surface of the scapula (supraspinous fossa), supraspinous fascia, and, to a certain extent, the subscapular fascia. Insertion: Greater tubercle of the humerus.

The extent of this muscle is evident only after removal of the loosely attached fleshy parts of the pectorals from its surface.

(f) The **subscapularis.** Origin: Entire medial surface of the scapula. Insertion: Lesser tubercle of the humerus.

(g) The **teres major.** Origin: Dorsal portion of the axillary border of the scapula. Insertion: In common with the latissimus dorsi on the anterior surface of the humerus.

(h) The **teres minor.** Origin: Ventral portion of the axillary border of the scapula. Insertion: Greater tubercle.

The muscle is closely associated with the infraspinatus but is separated from the teres major by the tendon of origin of the long head of the triceps.

(i) The **coracobrachialis.** Origin: Coracoid process. Insertion: Distal portion of the upper third of the humerus on its medial surface.

5. Muscles arising from the scapula and humerus and inserted on the proximal ends of the radius and ulna (extensors and flexors of the forearm), (Figs. 34, 35, 74).

A. **Extensor** (anconaeus) **group.** The muscles arise for the most part behind the axis of the humerus, and are inserted on the olecranon.

(a) The **extensor antibrachii parvus** (anconaeus quartus). Origin: Fascia of the medial surface of the humerus. Insertion: Medial surface of the olecranon.

The muscle should be divided, or detached from its origin, and reflected.

(b) The **anconaeus minimus** (epitrochleonanconaeus). Origin: Medial epicondyle of the humerus. Insertion: Medial surface of the olecranon

(c) The **triceps brachii.** Origin in three portions:
Caput longum (anconaeus longus): Ventral portion of the axillary border of the scapula.
Caput laterale (anconaeus lateralis): Greater tubercle and related portion of the lateral surface of the humerus.
Caput mediale (anconaeus medialis): Posterior surface of the humerus.

The three portions are almost separate muscles. Insertion on the olecranon.

B. **Flexor group.** The muscles arise in front of the axis of the humerus and are inserted on the radius and ulna in front of the elbow-joint.

(a) The **biceps brachii.** Origin: Anterior border of glenoid cavity. Insertion: Ventromedial surface of the ulna and medial surface of the radius. The muscle possesses only one head in the rabbit.

(b) The **brachialis.** Origin: Anterior and lateral surfaces of the humerus, divided unequally into a larger lateral and a smaller medial portion by the insertion tendons of the deltoideus and abductores muscles. Insertion: In common with the biceps.

6. Muscles arising from the distal end of the humerus and the radius and ulna and inserted on the hand (extensors and flexors of the hand and of the individual digits). The long insertion tendons pass through perforations of the **dorsal carpal** and **transverse** (ventral) **carpal ligaments** (Fig. 75).

A. **Extensor group.** The muscles have a general area of origin from the lateral epicondyle of the humerus and the antero-dorsal, or anterolateral surface of the radius and ulna. Insertion dorsal.

(a) The **extensor carpi radialis longus.** Origin: Lateral epicondyle. Insertion: Base of the second metacarpal.

(b) The **extensor carpi radialis brevis.** Origin: Lateral epicondyle. Insertion: Base of the third metacarpal. The muscle is partly fused with the foregoing one, and the tendons are closely associated on the wrist.

(c) The **abductor pollicis.** Origin: Anterolateral surface of the radius and ulna. Insertion: Base of the first meta-carpal. The muscle is partly concealed by (e). Its tendon forms a conspicuous cross with those of (a) and (b).

(d) The **extensor pollicis et indicis.** Origin: Anterolateral surface of the radius and ulna. Insertion: Ungual phalanx of the pollex and the head of the second metacarpal. Its tendon is the first of five in the centre of the carpus.

(e) The **extensor digitorum communis.** Origin: Lateral epicondyle and proximal end of the ulna. Insertion: By four tendons on all phalanges of the four lateral digits.

(f) The **extensor digiti quarti proprius.** Origin: Latera epicondyle. Insertion: Ungual phalanx of the fourth digit.

(g) The **extensor digiti quinti proprius.** Origin: Lateral epicondyle and lateral surface of the ulna. Insertion: Head of the fifth metacarpal and base of the first phalanx of this digit.

(h) The **extensor carpi ulnaris.** Origin: Lateral epicondyl and proximal portion of the lateral surface of the ulna Insercion: Base of the fifth metacarpal.

B. **Flexor group.** The muscles have a general area of origin from the medial epicondyle of the humerus and the posteroventral or posteromedial surface of the radius and ulna. Insertion volar.

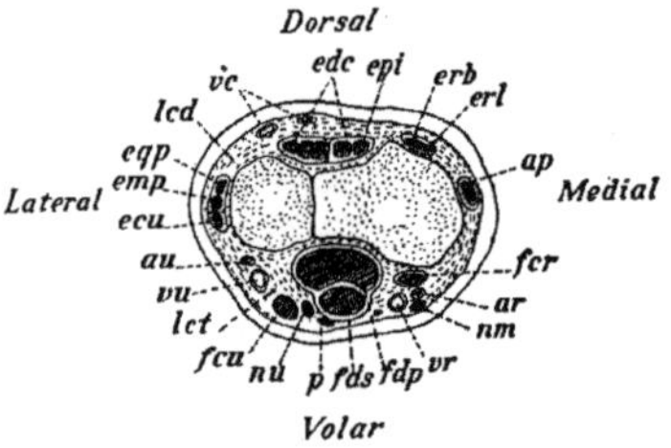

FIG. 75. Transverse section of the distal end of the forearm. Showing the relative positions of the muscle-tendons: ap, abductor pollicis; ar, radial artery; au, ulnar artery; ecu, extensor carpi ulnaris; edc, extensor digitorum communis; emp, extensor digiti quinti proprius; epi, extensor pollicis et indicis; eqp, extensor digiti quarti proprius; erb, extensor carpi radialis brevis; erl, extensor carpi radialis longus; fcr, flexor carpi radialis; fcu, flexor carpi ulnaris; fdp, flexor digitorum profundus; fds, flexor digitorum sublimis; lcd, dorsal carpal ligament; lct, transverse carpal ligament; nm, median nerve; nu, ulnar nerve; p, palmaris; r, radius; u, ulna; vc, cephalic vein; vr, radial vein; vu, ulnar vein.

(a) The **pronator teres.** Origin: Medial epicondyle. Insertion Ventral surface of th: radius.

(b) The **flexor carpi radialis.** Origin: Medial epicondyle. Insertion: Base of the second metacarpal.

(c) The **flexor digitorum sublimis.** Origin: In common with the ulnar portion of the profundus from the medial epicondyle; proximal portion of the ulna. Insertion: Bases of the second phalanges of the four lateral digits.

(d) The **palmaris.** Origin: Medial epicondyle. Insertion: Superficially on the volar fascia. This extremely slender muscle lies between the superficial portion of the profundus and the flexor carpi ulnaris.

(e) The **flexor digitorum profundus.** Origin in four portions:
Superficial portion: Medial epicondyle.
Radial portion: Ventral surface of the radius.

Middle portion: Ventral surface of the ulna.
Ulnar portion: Medial epicondyle in common with (c). Insertion: By five tendons on the bases of the ungual phalanges.

(f) The **flexor carpi ulnaris.** Origin: Medial epicondyle and medial surface of the olecranon, forming two short but separate heads. Insertion: Pisiform bone.

7. Muscles arising from the bones of the hand and inserted on the individual digits:

(a) The **flexor digiti quinti.** Origin: Pisiform bone and tendon sheath of the flexor digitorum profundus. Insertion: Sesamoid bones of the metacarpophalangeal joint of the fifth digit, extending to the ungual phalanx.

(b) The **lumbricales.** Origin: From the point of division of the tendon of the flexor digitorum profundus. Insertion: First phalanges of the second to fifth digits.

(c) The **interossei.** Origin: In pairs from the bases of the second to fifth metacarpals and related portions of the carpal bones. Insertion: Sesamoid bones of the metacarpophalangeal joints.

BLOODVESSELS AND NERVES OF THE ARM AND FOREARM.

The axillary artery gives rise to posterior and often anterior branches; the **circumflex arteries** to the head of the humerus and the **deep artery** which latter, arising in one or two branches and passing between the coraco-brachialis and teres muscles to the lateral surface of the shoulder, gives branches to the deltoideus and to the proximal ends of the lateral and long heads of the triceps. The deep artery continues on the lateral side of the medial head of the triceps and passes to the lateral head of the brachialis, near the elbow, as the **radial collateral artery.**

*The **brachial artery** (a. brachialis), the continuation of the axillary, passes distad on the medial surface of the arm between the biceps and the anconaeus medialis. Crossing to the anterior surface of its distal extremity, it passes beneath the head of the pronator teres to the medial surface of the radius, dividing at this

point—a short distance in front of the elbow—into the radial and ulnar arteries. Its chief branches on the arm are the **ulnar collateral arteries** (superior, middle, and inferior) to the muscles and the elbow joint.

The **radial artery** (a. radialis) passes distad on the ventromedial border of the radius, in company with the median nerve, lying at first between the flexor carpi radialis and the radial portion of the flexor digitorum profundus. Toward the distal end of the radius it crosses the ventral surface of the tendon of the flexor carpi radialis, and appears in a superficial position on the medial border of the carpus. It reaches the volar surface of the hand after passing obliquely across the tendon of the flexor digitorum sublimis.

The **ulnar artery** (a. ulnaris) crosses the ulna obliquely from its origin, reaching in this way the lateral border of the flexor carpi ulnaris, along which it passes to the end of the forearm and to the ulnar side of the pisiform bone. It passes to the ventral surface of the fifth digit, and then turns across the hand, forming the volar arch.

The single **brachial vein (v. brachialis) accompanies the brachial artery and lies behind it. It is formed in front of the elbow by the union of two vessels, the **radial** and **ulnar veins**, which accompany the corresponding arteries and join one another at the point of separation of the latter. The radial vein anastomoses with the radial portion of the cephalic at a point distal to the middle of the forearm.

The **cephalic vein** (v. cephalica) is a large superficial vessel appearing on the dorsal surface of the forearm. From the radial side of the latter it receives a large tributary which anastomoses with the radial vein. It is accompanied by branches of the superficial ramus of the radial nerve. It passes to the front of the arm across the angle of the elbow, lying at first on the anterior margin of the lateral head of the brachialis, and afterwards on the lateral surface of the arm between the abductor brachii superior and the anconaeus lateralis. It disappears from this surface in the triangular space enclosed by these muscles and the insertion of the levator scapulae major, receiving at this point a large tributary from the shoulder. It appears on the medial surface of the shoulder at

the distal end of the axillary border of the scapula between the teres major and subscapularis, entering the axillary vein at about the same place as the subscapular vein, or in common with the latter.

***The **radial nerve** (n. radialis) passes behind the brachial artery to the posterior surface of the humerus. It perforates the anconaeus medialis, appearing afterwards on the lateral side of the brachialis in company with the collateral radial artery. A **superficial ramus**, given off on the distal portion of the arm, accompanies the cephalic vein: it passes along the surface of the extensor carpi radialis, dividing into branches for the dorsum of the hand. The remaining portion is chiefly distributed as the **ramus profundus** to the extensor muscles of the forearm.

The **median nerve** (n. medianus) passes distad along the medial surface of the arm, lying at first in front of the brachial artery and then on its medial side. It accompanies the brachial artery, passing beneath the head of the pronator teres, and then traverses the forearm, in company with the radial artery, to the volar surface of the hand.

The **ulnar nerve** (n. ulnaris) lies behind the brachial artery. Toward the distal extremity of the humerus it accompanies the inferior ulnar collateral artery. It passes from the medial surface of the elbow, between the anconaeus minimus and the base of the olecranon, to the dorsal surface of the olecranon head of the flexor carpi ulnaris, and then crosses the ulna obliquely, in company with the ulnar artery, to the lateral border of the muscle and along it to the insertion tendon. At the wrist it crosses the dorsal surface of the tendon, and passing between the tendon of the sublimis and the pisiform bone, reaches the volar surface of the hand.

IX. THE POSTERIOR LIMB.

Dissection on the side opposite to that of injection. The dissection is largely a muscular one; to be conducted in the same way as in the anterior limb. The corresponding muscle groups should be compared with respect to the difference in orientation of the equivalent segments.

1. Muscles arising from the ventral surface of the posterior thoracic and lumbar vertebrae and inserted on the pelvic girdle, or on the lesser trochanter of the femur. These muscles are chiefly distinguished by their vertebral position, on account of which and the fixed condition of the pelvic girdle they combine the characters of vertebral and appendicular muscles.

(a) The **psoas minor.** Origin: Bodies of the four posterior lumbar vertebrae. Insertion: Pecten of the pubis. The flat, pointed tendon forms a square cross with a dorsal (sacral) continuation of the inguinal ligament, which is stretched transversely from the middle of the inguinal ligament to the centre of the body of the first sacral vertebra, and on which the tendon is also inserted.

It is necessary to divide the inguinal ligament and reflect its sacral continuation, together with the tendon of the psoas minor.

(b) The **psoas major.** Origin: Internal surfaces of the bases of the last three ribs and bodies of the corresponding thoracic vertebrae; also the lumbar vertebrae. Insertion: Lesser trochanter.

(c) The **iliacus.** Origin: Bodies of the last lumbar and first sacral vertebrae, extending to the sacroiliac union. Insertion: With the psoas major on the lesser trochanter. The two muscles together form the **iliopsoas.**

The **lumbar portion** of the **lumbosacral plexus,** beginning with the **fourth lumbar** nerve, lies on the ventral surface of the psoas major and between the latter and the iliacus. The **fifth** lumbar is the chief root of the **femoral nerve** (p. 225), the trunk of which may be traced from a position between the two muscles distad over the dorsal surface of the inguinal ligament to the medial surface of the thigh. The remaining four nerves crossing the dorsal body-wall obliquely are the twelfth thoracic and first three lumbar.

The psoas major should be freed at its lateral margin and turned toward the middle line, the fourth lumbar nerve being divided.

(d) The **quadratus lumborum.** Origin: Bodies of the posterior five thoracic vertebrae and the bases of the corresponding five ribs; bodies and transverse processes of the lumbar vertebrae. Insertion: Triangular processes of six lumbar vertebrae and the posterior ventral angle of the iliac wing, together with the adjacent portion of its medial surface.

2. Muscles arising from the pelvic girdle and sacrum and inserted on the femur, for the most part at its proximal extremity.

The muscles of this group enclose the proximal portion of the femur on its lateral, posterior, and medial sides. They are partly covered by the flexors of Group 3, namely, the biceps, sartorius, and gracilis (pp. 222, 223), which must be divided. To begin the dissection, see directions for removing the biceps (p. 223). The position of the sciatic vein (p. 225) should be noted.

Dissection on the lateral surface posteriorly.

(a) The **glutaeus maximus.** Origin in two fleshy portions, joined by an aponeurosis:

First portion: Fascia covering the sacrum in its entire length. This portion is triangular in shape, and is covered posteriorly by the first head of the biceps.

Second portion: Anteroventral border of the iliac wing, fused with the tensor fasciae latae and the first head of the rectus femoris; also from the dorsal border and lumbar fascia through the aponeurosis mentioned above.

Insertion: Third trochanter. The axis of the first portion is transverse, that of the second horizontal.

Both portions of the muscle should be divided. The sciatic nerve and artery are then exposed.

(b) The **glutaeus medius.** Anterior border of the wing of the ilium and the iliac crest and fascia of the first two sacral vertebrae. Insertion: Greater trochanter. Some of the fibres pass around the medial surface of the tip of the greater trochanter and are inserted in the lateral wall of the trochanteric fossa.

The muscle should be divided.

(c) The **glutaeus minimus.** Origin: Entire lateral surface of the body and wing of the ilium. Insertion: Greater trochanter.

Remove the entire muscle. The piriformis is in contact with its dorsal margin posteriorly.

(d) The **tensor fasciae latae.** Origin: Anterior portion of the ventral border of the wing of the ilium. Insertion: Broad

fascia of the lateral surface of the thigh. The muscle is fused with the first head of the rectus femoris in front, and with the second portion of the glutaeus maximus behind.

(e) The **piriformis.** Origin: Lateral portions of the second and third sacral vertebrae. Insertion: Tip of the great trochanter.

The muscle should be divided, care being taken to avoid injury to the nerves and bloodvessels beneath it.

(f) The **gemellus superior.** Origin: Tendinous from the ischial spine and fleshy from the body of the ischium immediately in front of it. Insertion: Lateral wall of the trochanteric fossa.

The muscle extending from the ischial spine to the sacrum is the abductor caudae anterior (p. 270).

(g) The **obturator internus.** Origin: Internal extent of the obturator foramen. Insertion: Trochanteric fossa. The muscle passes over the ischium in the lesser sciatic notch, only its insertion portion appearing from the lateral surface. To see its origin, reflect the tendon through the lesser sciatic notch and examine the muscle from the internal surface of the pelvis.

(h) The **gemellus inferior.** Origin: Posterior portion of the superior ramus of the ischium and the ischial tuberosity. Insertion: Trochanteric fossa.

(i) The **quadratus femoris.** Origin: Ventral surface of the ischial tuberosity and the superior ramus of the ischium immediately in front of it. Insertion: The superficial fibres are inserted on and below the third trochanter, the remaining ones below the trochanteric fossa.

(j) The **obturator externus:** Origin: External extent of the obturator foramen. Insertion: Trochanteric fossa. The muscle is largely concealed from this surface, but may be fully displayed by the division of the pectineus and adductores brevis and longus.

Dissection on the medial surface posteriorly, after division of the gracilis.

(k) The **pectineus.** Origin: Pecten of the pubis. Insertion Immediately below the lesser trochanter.

(l) The **adductor brevis.** Origin: Anterior portion of the symphysis pubis. Insertion: Below the pectineus.

(m) The **adductor longus.** Origin: Posterior portion of the symphysis and inferior ramus of the ischium. Insertion: Posterior surface of the shaft of the femur to its distal third.

(n) The **adductor magnus.** Origin: Ventral surface of the ischial tuberosity. Insertion: Medial surface of the distal end of the femur, extending to the medial condyle of the tibia.

3. Muscles arising from the pelvic girdle and the femur and inserted on the proximal portions of the tibia and fibula (extensors and flexors of the leg). (Fig. 76).

A. **Extensor group (quadriceps femoris).** The muscles lie for the most part in front of the axis of the femur. They have a common insertion on the tibial tuberosity through the patella and the patellar ligament.

(a) The **rectus femoris.** Origin in two portions:

First portion: Superior anterior spine, fused with the tensor fasciae latae, and ventral border of the iliac wing.

Second portion: By a stout round tendon from the inferior anterior spine, immediately in front of the acetabulum. This part is cylindrical in shape and is almost a separate muscle.

The two portions of the muscle should be divided.

(b) The **vastus lateralis.** Origin: Anterior surface of the great trochanter and the lateral intermuscular ligament.

The muscle should be divided.

(c) The **vastus intermedius.** Origin in two portions:

First portion: Great trochanter, below the origin of the vastus lateralis.

Second portion: Anterior surface of the femur.

(d) The **vastus medialis.** Origin: Medially, at the base of the collum femoris and adjacent portion of the shaft.

Common insertion (a-d): Tibial tuberosity.

B. **Flexor group** (hamstring muscles). With the exception of the sartorius, the muscles lie behind the axis of the femur, and are

inserted on the medial and lateral surfaces of the knee-joint and the corresponding proximal portions of the leg. They form the boundaries of the popliteal fossa.

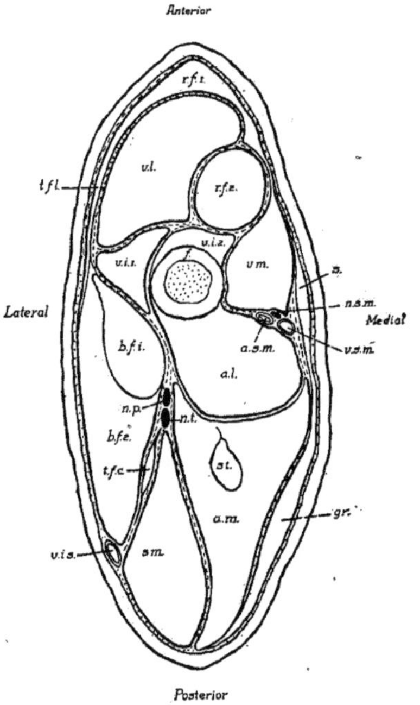

FIG. 76. Transverse section through the middle of the thigh: a.l., adductor longus; a.m., adductor magnus; b.f.1 and b.f.2, first and second heads of the biceps femoris; f., femur; gr., gracilis; n.p., peroneal nerve; n.s.m, greater saphenous nerve; n.t., tibial nerve; r.f. 1 and r.f. 2, first and second heads of the rectus femoris; s, sartorius; sm., semimembranosus; st., semitendinosus; t.f.c., tensor fasciae cruris; t.f.l., tensor fasciae latae; v.i.1 and v.i. 2, first and second heads of the vastus intermedius; v.is. sciatic vein; v.l., vastuo lateralis; v.m., vastus medialis; v.s.m., great saphenousvein.

(a) The **sartorius.** Origin: Posterior portion of the inguinal ligament, especially its sacral extension. Insertion: Medial condyle of the tibia. This is an extremely thin and narrow band of fibres, lying on the more anterior portion of the medial surface of the thigh. The muscle is fused distally with the gracilis. It has the position of a rotator muscle, and is only a flexor through its connection with the latter.

(b) The **gracilis.** Origin: The entire extent of the pubic symphysis. Insertion: Through a broad tendinous expansion ending in the fascia of the proximal portion of the medial surface of the thigh. The muscle forms a broad, comparatively thin sheet, covering the posterior portion of the medial surface of the thigh. Its insertion tendon is perforated

by the great saphenous artery and vein and the greater saphenous nerve.

The sartorius and gracilis should be raised from the surface and divided.

(c) The **biceps femoris**: Origin in two portions:

First portion (caput breve): Spinous processes of three posterior sacral and three anterior caudal vertebrae. This portion is triangular in shape, the distal end, or apex of the triangle, passing into a thin flat tendon which is inserted on the lateral margin of the patella.

Second portion (caput longum): Dorsal surface of the ischial tuberosity, fused with the adductor magnus, and the lateral process, fused with the second, or deep portion of the semimembranosus. This portion is also triangular, the base being distal and providing a broad insertion on the fascia of the proximal third of the lateral surface of the leg. The first, or superficial head of the semimembranosus, covers this portion at its origin, which is also crossed by the sciatic vein.

To free the biceps and divide it: Incision along the aponeurotic line joining the tip of the great trochanter with the sacrum; also distad along the intermuscular septum of the lateral surface of the thigh from the great trochanter to the knee. An incision along the first portion of the sciatic vein will free the superficial head of the semimembranosus from the biceps. Raise the biceps slightly by working the handle of the scalpel under its distal portion. Divide the muscle from its posterior margin, being careful not to injure the slender tensor fasciae cruris muscle and the branches of the sciatic nerve which lie beneath it. The first head of the biceps may be separated and reflected in order to expose the short muscles of the thigh.

(d) The **tensor fasciae cruris**. Origin: By a long, slender tendon, from the transverse process of the fourth sacral vertebra. Insertion: Lateral fascia of the leg. This slender muscular slip underlies the biceps femoris.

(e) The **semimembranosus.** Origin in two portions:
First (superficial) **portion:** Fascia covering the first head of the biceps.
Second (deep) **portion:** Lateral process of the ischial tuberosity. Insertion: In common with the gracilis on the fascia of the proximal portion of the medial surface of the leg. This fascia is contracted into two ligaments, one of which carries the insertion of the muscle to the distal end of the tibial tuberosity, the other to the distal end of the leg, where it joins the tendon of the heel (tendo calcaneus).

(f) The **semitendinosus.** Origin: Ischial tuberosity. Insertion: Medial condyle of the tibia. The muscle is completely enclosed by the adductor magnus, which must be split to expose it.

BLOODVESSELS AND NERVES OF THE THIGH.

*The **femoral artery** (a. femoralis) traverses the medial surface of the thigh, beginning at the dorsal side of the inguinal ligament, where it continues the external iliac artery. Immediately distal to the inguinal ligament it gives off the **deep artery** of the thigh (a. profunda femoris). The latter passes to the dorsal side of the pectineus and adductor brevis muscles and is distributed to the posterior proximal portion of limb, chiefly to the adductores longus and magnus. A second branch, the lateral circumflex artery (a. circumflexa femoris lateralis), is given off from the anterior wall. It passes between the second head of the rectus femoris and vastus lateralis, on the one hand, and the two portions of the vastus intermedius, on the other. It supplies various parts of the quadriceps femoris group. A third branch of the femoral, the **superficial epigastric artery** (a. epigastrica superficialis), given off medially, and passing to the abdominal wall, has been divided (p. 178). At the beginning of the distal third of the thigh, a small branch, the **a. genu suprema,** passes over the medial condyle of the femur to the knee-joint, and at about the point of origin of this vessel a large branch, the **great saphenous artery** (a. saphena magna), arises from the posterior wall. It passes across the medial surface of the distal end of the adductor longus, and through the tendon of the gracilis, to the medial

surface of the leg. The femoral artery passes between the adductores longus and magnus, continuing as the **popliteal artery** (a. poplitea).

The hypogastric artery (p. 203) appears in the greater sciatic notch, continuing thence as the **sciatic artery** (a. ischiadica). The vessel passes backward to the dorsal surface of the superior ramus of the ischium, where it divides into **lateral caudal** and **internal pudendal** branches. Its smaller branches are distributed to the glutaei and biceps femoris muscles.

The **femoral vein (v. femoralis) traverses the medial surface of the thigh in company with the femoral artery. It begins at the proximal end of the lower third of the thigh as a continuation of the **popliteal vein** (v. poplitea), which accompanies the corresponding artery. Its tributaries comprise the **great saphenous, superficial epigastric, lateral circumflex,** and **profunda femoris** veins.

The **sciatic vein** (v. ischiadica) traverses the lateral surface of the thigh near its posterior margin, lying at first between the biceps and semimembranosus and afterwards on the posterior margin of the former. At the dorsal border of the ischium, in front of the ischial tuberosity, it receives the **lateral caudal** and **internal pudendal** veins.

***The **femoral nerve** (n. femoralis) arises from the lumbosacral plexus, chiefly from the fifth lumbar. Its position between the psoas major and iliacus muscles has already been noted (p. 218). Immediately beyond the inguinal ligament it divides into two portions, one of which is distributed to the muscles of the anterior side of the thigh, while the other, the **greater saphenous nerve** (n. saphenus major), passes to the medial surface of the thigh and leg in company first with the femoral artery and afterwards with the great saphenous artery.

The **sciatic nerve** (n. ischiadicus), formed chiefly from the seventh lumbar and first sacral nerves, appears laterally in the greater sciatic notch. It passes backward beneath the piriformis muscle, and then turns distad to the thigh, where it lies on the lateral surfaces of the adductores magnus and longus. It distributes branches to the posterior musculature of the thigh. In the proximal portion of the thigh it divides into two chief branches, which are closely associated as far as the knee. The anterior

branch is the **peroneal nerve** (n. peronaeus), the posterior branch the **tibial nerve** (n. tibialis). The lesser saphenous nerve (p. 231) is a small branch given off from the tibial above the knee-joint.

For the origin of this and related nerves see p. 232.

The **superior gluteal nerve** (n. glutaeus superior) appears in the greater sciatic notch, leaving the sciatic close to the inferior posterior spine of the ilium. It passes between the glutaeus minimus and the lateral surface of the ilium, ending in the tensor fasciae latae. Its branches are distributed to the glutaei medius and minimus and the piriformis muscles.

The **inferior gluteal nerve** (n. glutaeus inferior) perforates the posterior portion of the piriformis, and is distributed to the glutaeus maximus.

The **posterior cutaneous nerve** (n. cutaneus femoris posterior) accompanies the hypogastric artery backward to the ischial tuberosity, where it turns to the posterior margin of the thigh and the medial surface of the sciatic vein, ending in branches to the skin.

The **pudendal nerve** (n. pudendus) accompanies the sciatic artery and afterwards the internal pudendal to the penis or clitoris.

In preparation for the muscular dissection of the leg, the insertion tendons of the biceps femoris, tensor fasciae cruris, gracilis, and semimembranosus muscles should be removed from about the knee-joint. The adductor magnus may be detached from the medial condyle of the femur, but the popliteal vessels must be kept intact. The superficial blood-vessels of the leg should be noted, since it is necessary to clear them away in separating the muscles. They include, medially, the great saphenous artery and vein, and laterally the branches of the small saphenous artery to the insertion portions of the muscles of the thigh, and its continuation on the postero-lateral border of the leg; and the sciatic vein, together with its continuation, the anterior tibial vein, and the accessory small saphenous vein (p. 231). The tibial and peroneal nerves may be cut, after first noting their position.

4. Muscles arising from the medial and lateral condyles of the femur and from the proximal portions of the tibia and fibula, including the tibial condyles; inserted on the foot. The group

includes the typical extensors and flexors of the foot, together with the peronaei muscles, which individually are extensors and flexors, but collectively have the relation of lateral tractors (Fig. 77).

A. **Extensor group.** Muscles occupying an anterior position on the leg and inserted on the dorsum of the foot.

(a) The **extensor hallucis longus.** Origin: Middle portion of the anteromedial surface of the tibia and from the medial condyle behind the tibial collateral ligament. Insertion: The tendon passes around the medial malleolus of the tibia and beneath the base of the second (first functional) metatarsal, continuing to its dorsal surface and uniting with the first tendon of the extensor digitorum longus.

The posterior tibial artery, the continuation of the great saphenous, and the tibial nerve accompany the tendon in the malleolar groove.

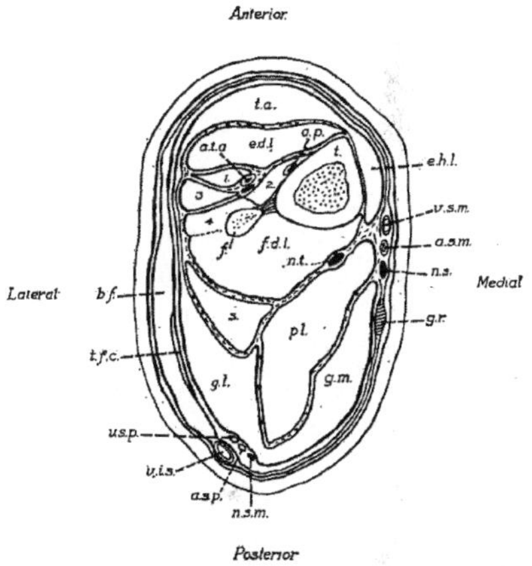

FIG. 77. Transverse section of the proximal portion of the leg: a.s.m., great saphenous artery; a.s.p., small saphenous artery; a.p., a.t.a., anterior tibial artery; b.f., biceps femoris (insertion); e.d.l., extensor digitorum longus; e.h.l., extensor hallucis longus; f., fibula; f.d.l., flexor digitorum longus; g.l., and g.m., lateral and medial heads of the gastrocnemius; gr., gracilis (insertion tendon); n.s., greater saphenous nerve; n.s.m., lesser saphenous nerve; n.t., tibial nerve; pl., plantaris; s., soleus; t., tibia; t.a., tibialis anterior; t.f.c., tensor fasciae cruris (insertion); v.is., sciatic vein; v.s.m., great saphenous vein; v.s.p., small saphenous vein; 1-4, the personaei (primus-quartus).

(b) The **tibialis anterior.** Origin: Lateral condyle of the tibia and corresponding surface of the tibial tuberosity. Insertion: Base of the second metatarsal. The tendon passes beneath the obliquely placed crural ligament of the lower portion of the leg.

The muscle should be divided and its head reflected.

The anterior tibial artery and vein, and the peroneal nerve appear on the anterior surface of the tibia at the lateral side of the crural ligament.

(c) The **extensor digitorum longus.** Origin: By a flattened tendon from the lateral portion of the patellar surface of the femur. This tendon passes through the capsule of the knee-joint, and the fleshy portion of the muscle lies on the antero-lateral surface of the tibia. Insertion: The distal tendon passes beneath the crural ligament, then beneath the cruciate ligament of the dorsum of the foot, dividing into four portions for insertion on all the phalanges of the digits.

The muscle may be displaced by dividing the crural ligament.

The branches of the anteror tibial artery lie behind this muscle, one in a medial position, in contact with the tibia, the other on the peronaei muscles in company with the peroneal nerve.

B. **Peronaeus group.** These muscles arise from the lateral surface of the leg, and are inserted on all surfaces at the lateral side of the foot. The insertion tendons reach the foot from beneath the lateral malleolus. The muscles can be separated after the tendons are released from this position.

(a) The **peronaeus longus** (p. primus). Origin: Lateral condyle of the tibia and head of the fibula. Insertion: End of the reduced first metatarsal. The tendon crosses the plantar surface of the foot, passing around the distal end of the cuboid bone.

The muscle should be divided.

(b) The **peronaeus brevis** (p. secundus). Origin: Lateral condyle of the tibia and corresponding surface of the shaft; also the crural interosseous ligament joining the tibia and fibula. Insertion: Tuberosity of the base of the fifth metatarsal.

(c) The **peronaeus tertius.** Origin: The head of the fibula and the crural interosseous ligament, fused with the extensor digitorum longus. Insertion: Head of the fifth metatarsal, and distally, united with the tendon of the extensor digitorum longus, on the phalanges of this digit.

(d) The **peronaeus quartus.** Origin: The fibula and the interosseous ligament, fused with the peronaeus brevis and with the fiexor digitorum longus. Insertion: Head of the fourth metatarsal.

C. **Flexor group.** The muscles arise from the medial and lateral condyles of tibia and femur (the flexor digitorum sublimis from the posterior surface of the tibia). They lie behind the axis of the tibia, and are inserted both on the heel and on the plantar surface of the foot.

(a) The **triceps surae** comprises:

(1) The **gastrocnemius.** Origin: in two portions:

Caput laterale: Lateral condyles of tibia and femur and related femoral sesamoid.

Caput mediale: Medial condyle of the femur and its sesamoid.

(2) The **soleus:** Origin: By a strong tendon from the head of the fibula.

Insertion: Through the **Achilles' tendon** (tendo calcaneus). The latter passes over the posterior end of the tuber calcanei, and is attached to its ventral surface. The tendon is covered by that of the plantaris muscle.

The small saphenous artery and vein lie at the posterior margin of the lateral head of the gastrocnemius in company with the lesser saphenous nerve.

(b) The **plantaris.** Origin: Lateral condyle of the femur and associated sesamoid. Insertion: The tendon passes over the heel to the plantar surface of the foot, and divides into four parts for insertion on the second phalanges of the four developed digits.

The two muscles should be divided.

(c) The **popliteus.** Origin: Lateral condyle of the femur: The tendon passes through the capsule of the knee-joint. The muscle contains the tibial sesamoid. It crosses the posterior surface of the tibia obliquely, and is inserted on the proximal portion of its posteromedial angle.

(d) The **flexor digitorum longus.** Origin: Lateral condyle of the tibia and head of the fibula, extending to the posterior surface of the interosseous ligament and associated portions of the tibia and fibula. Insertion: The tendon passes beneath the sustentaculum tali, reaching the plantar surface of the foot, where it is partly covered by the plantaris tendon. It divides into four parts for insertion on the ungual phalanges of the four developed digits.

The tibial nerve lies on the medial surface of the head of the plantaris and afterwards on the medial surfaces of the popliteus and flexor digitorum longus.

5. Muscles arising from the foot and inserted on the individual digits.

(a) The **lumbricales.** Origin: Tendon of the flexor digitorum longus. Insertion: Medial surfaces of the first phalanges of the three lateral digits.

(b) The **interossei** (metatarsi). Origin: From the dorsal portion of the tendon-sheath of the flexor digitorum longus. Insertion: Heads of the four metatarsals.

VESSELS AND NERVES OF THE LEG AND FOOT.

*The **great saphenous artery** passes distad on the medial surface of the leg, and is continued as the **posterior tibial artery** (a. tibialis posterior) around the medial malleolus to the plantar surface of the foot. Above the ankle-joint it gives off the **malleolar artery** (a. malleolaris) to the posterior surface of the distal end of the tibiofibula.

The **popliteal artery,** the continuation of the femoral, passes between the medial head of the gastrocnemius on the one hand and the lateral head and the plantaris on the other, reaching the anterior surface of the popliteus, and afterwards the anterior surfaces of the tibia and fibula by passing between their proximal ends. It distributes branches to the muscles about the knee-joint, including a branch to the distal portion of the vastus lateralis, which is given off at about the point of origin of the small saphenous artery. It then continues as the **anterior tibial artery.** The vessel appears in front of the interosseous ligament of the leg and of the peronaeus brevis, and continues to the dorsum of the foot after passing beneath the crural ligament. A large branch, given off in the upper part of the leg also reaches the dorsum of the foot from a more lateral position.

The **small saphenous artery** (a. saphena parva) appears on the proximal portion of the posterolateral margin of the leg, running along the border of the lateral head of the gastrocnemius in company with the corresponding vein and the lesser saphenous nerve.

The **great saphenous vein (v. saphena magna), a large tributary of the femoral, accompanies the corresponding artery, and the greater saphenous nerve, and is continued as the **posterior tibial vein** (v. tibialis posterior), to the plantar surface of the foot.

The **popliteal vein,** the root of the femoral, accompanies the corresponding artery in the popliteal fossa. It receives the **small saphenous vein** (v. saphena parva) from the posterior margin of the lateral head of the gastrocnemius.

The **sciatic vein** is continued on the lateral surface of the leg as the **anterior tibial vein** (v. tibialis anterior). It receives the **accessory small saphenous vein** (v. saphena parva accessoria) from the posterior surface, and is continued to the dorsum of the foot passing to the fibular side of the crural ligament.

***The **greater saphenous nerve,** the posterior branch of the femoral nerve, accompanies first the femoral artery and afterwards the great saphenous artery, passing distad to the medial surface of the leg.

The **tibial nerve,** the posterior division of the sciatic, passes between the medial and lateral heads of the gastrocnemius to the medial surface of the head of the plantaris. It traverses the leg, lying on the medial surface first of the popliteus and afterwards of the flexor digitorum longus, and passing beneath the medial malleolus reaches the plantar surface of the foot. In the proximal portion of the leg it distributes muscular branches to the flexor group.

The **lesser saphenous nerve** (n. saphenus minor) accompanies the small saphenous artery and vein on the posterior margin of the lateral head of the gastrocnemius.

The **peroneal nerve,** the anterior division of the sciatic, passes distad, lying at first between the insertion of the biceps and the lateral head of the gastrocnemius, and thus appearing on the surface after the removal of the former. It perforates the anterior portion of the lateral head of the gastrocnemius and afterwards the fused heads of the peronaeus tertius and flexor digitorum longus, traversing the leg at first behind the peronaeus longus and then around its medial margin to the front of its tendon. It passes to the fibular side of the crural ligament and thence to the dorsum of the foot. The nerve distributes branches to the tibialis anterior and extensor digitorum longus.

THE LUMBOSACRAL PLEXUS.

The structure of the lumbosacral plexus may be examined by breaking away the ventral portion of the pelvis, or by dividing the sacroiliac articulation in such a way that the two sides of the pelvis may be pressed apart, the ventral or pelvic face of the sacrum being thus exposed. The posterior portions of the psoas and iliacus muscles may be picked away with the forceps, and the abductor caudae anterior muscle (p. 270) may be detached from its origin on the ischial spine.

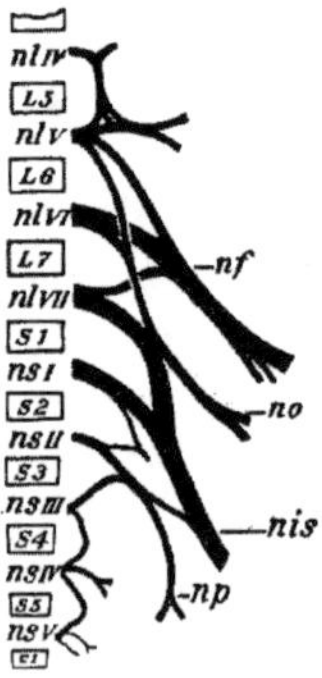

FIG. 78. Plan of the lumbosacral plexus: L,S,C, lumbar, sacral, and caudal vertebrae; nl, ns, corresponding spinal nerves; nf, femoral nerve; no, obturator nerve; nis, sciatic nerve; np, pudendal nerve.

The **lumbosacral plexus** (plexus lumbosacralis) is formed by the ventral roots of the four posterior lumbar and four sacral spinal nerves (Fig. 78). It is divisible into a **lumbar plexus** (plexus lumbalis), from which arises the femoral nerve, and a **sacral plexus** (plexus sacralis), from which arises the sciatic nerve.

The **femoral nerve** is formed from the fifth, sixth, and seventh lumbar, especially from the loop connecting the fifth and sixth (ansa lumbalis II). The **obturator nerve** (n. obturatorius), which accompanies the obturator artery, is formed from the fifth, sixth and seventh lumbars but chiefly from the sixth, and is distributed to the obturatores, adductores, and gracilis muscles.

The **sciatic nerve,** together with the superior and inferior gluteal nerves, arises chiefly from the loop connecting the last lumbar and first sacral nerves (ansa lumbalis III).

The internal pudendal nerve is formed from the loop connecting the second and third sacral nerves (ansa sacralis II), but chiefly from the second.

THE ARTICULATIONS OF THE POSTERIOR LIMB.

The more perfect development and larger size of the joints of the posterior limb make them much more favorable for examination than the corresponding parts of the anterior limb.

The muscular attachments should be removed from about the articular capsules and the structures examined as follows:

A. The **hip-joint** (articulatio coxae) is an **enarthrosis,** formed by the head of the femur with the parts of the ischium, ilium, and the os acetabuli enclosing the acetabulum, together with the **articular capsule** (capsula articularis) and accessory ligaments.

The articular capsule extends from the acetabular margin to the proximal end of the neck of the femur. It is strongest on its dorsal side, but is especially thickened at three points, forming the **ischiocapsular** (dorsal), **iliofemoral** (anterior), and **pubocapsular** (ventral) **ligaments.**

By dividing the capsule, the contents of the joint and the smooth articular surfaces may be examined; also the attachment of the head of the femur to the wall of the acetabular fossa through the **round ligament** (lig. teres femoris). The **glenoid lip** (labrum glenoidale) is the ring of fibrocartilage surrounding the ma gin of the acetabulum and connecting with the articular capsule.

B. The **knee-joint** (articulatio genu) is a hinge-joint or **ginglymus** with a slight spiral trend. It is formed by the articular surfaces of the medial and lateral condyles of the femur and tibia, with the associated articular capsule, ligaments, and inarticular fibrocartilages (see section, Fig. 13).

The **tibial collateral ligament** (lig. collaterale tibiale) is a stout band of connective tissue stretching fr m the medial condyle of the femur to the posteromedial angle of the medial condyle of the tibia.

The **fibular collateral ligament** is a similar structure connecting the lateral condyle of the femur with the anterior surface of the head of the fibula.

The sesamoid bones of the popliteal region have articular surfaces taking part in the formation of the joint. That on the medial condyle of the femur is contained in the medial head of the gastrocnemius, that on the lateral condyle of this bone in the lateral head of the gastrocnemius and the plantaris, and that on the lateral tibia condyle in the popliteus.

The common tendon of the quadriceps femoris, the patella, and the patellar ligament are associated with the capsule, forming the anterior wall of the joint.

Between the apposed surfaces of the condyles, in the interior of the joint, there are two short, **cruciate** ligaments and two thin plates of fibrocartilage, the **medial** and **lateral menisci.** The **anterior cruciate ligament** (lig. cruciatum anterius) passes from the lateral wall of the intercondyloid fossa of the femur to the anterior end of the intercondyloid eminence of the tibia. The **posterior cruciate ligament** passes from the medial wall of the intercondyloid fossa of the femur to the posterior intercondyloid fossa of the tibia. The **medial meniscus** (meniscus medialis), a thin crescentic plate of fibro-cartilage, lies on the articular surface of the medial condyle of the tibia, and is connected by ligament with the anterior and posterior intercondyloid fossae of the bone. The larger, **lateral meniscus** lies on the lateral condyle of the tibia, and is attached by ligament anteriorly to the medial portion of the articular surface, and posteriorly to the medial wall of the intercondyloid fossa of the femur. The tendon of origin of the extensor digitorum communis traverses the anterior part of the joint on its way from the patellar surface of the femur to the front of the leg.

The **interosseous ligament** of the leg (lig. interosseum cruris) forms an almost complete sheet connecting the uncoalesced portions of tibis and fibula.

C. The **ankle-joint** (articulatio talocruralis) is a ginglymus with a considerable amount of spiral torsion. The articulating surfaces are chiefly formed by the tibia and talus, but also by the fibular side of the tibiofibula and the calcaneus. On the tibial side the **calcaneotibial ligament** (lig. calcaneotibiale) connects the medial malleolus with the sustentaculum tali. On the fibular side the **calcaneofibular ligament** (lig. calcaneofibulare) connects the posterior portion of the groove for the peronaei muscles forwards with the lateral surface of the calcaneus, and a second ligament extends from the anterior margin of the groove backward to the lateral surface of the calcaneus. The **tibionavicular ligament** (lig. tibionaviculare) connects the anterior surface of the distal end of the tibia with the dorsal surface of the navicular bone. The joint contains in its interior a short strong ligament connecting the medial side of the lateral malleolus with the lateral and ventral surfaces of the trochlea tali.

X. THE HEAD AND NECK.

This dissection includes the various structures of the region, with the exception of the cervical and occipital musculature and the central nervous system, which are treated in the succeeding parts, and the special musculature of the ear, which has been omitted.

To begin the dissection, the skin, which has already been divided to the mandibular symphysis and partly reflected, should be separated from the underlying platysma along the side of the head, and reflected until the surface is clear to a point near the dorsal median line of the skull. The insertion of the platysma on the cheek, and that of its special portion, the depressor conchae posterior, on the base of the ear, should be noted. A second band of muscle, similar in its relation to the platsyma, arises from the lateral border of the mandible, immediately in front of the masseter muscle, and is inserted into the base of the ear. This is the **depressor conchae** (parotideoauricularis) **anterior.**

In removing the skin of the upper and lower eyelids, two muscles, the **orbicularis oculi** and the **depressor palpebrae inferioris,** may be observed. The former is a somewhat circular band of fibres enclosing both eyelids, the fibres lying directly on the skin, and being concentrated at the anterior and posterior angles. The latter is a very slender muscle arising from the zygomatic arch and inserted into the skin of the lower eyelid. The corresponding levator palpebrae superioris arises from the orbital wall, and is here concealed by the projecting supraorbital process.

1. On the lateral surface of the head the following structures may be made out without further dissection:

(a) The **masseter muscle.** Origin: The zygomatic arch; tendinous from its anterior angle. Insertion: Lateral surface of the angle of the mandible.

(b) The **parotid gland** (gl. parotis), a diffuse, white or brownish gland lying immediately behind the angle of the mandible. Its **duct** (d. parotideus) crosses the masseter and perforating the mucous membrane of the cheek opens into the oral cavity.

(c) The chief part of the **seventh cranial** or **facial nerve** (n. facialis) appears in the anterior portion of the parotid gland,

its branches crossing the masseter. They are distributed as motor nerves to the cutaneous muscles of the face, including the platysma.

(d) The **external maxillary artery** (a. maxillaris externa) appears at the ventral border of the mandible immediately in front of the masseter. It passes upward to the anterior margin of the eye, ending in the **angular artery** (a. angularis). Its chief branches to the anterior portion of the face are: (1) the **submental artery** (a. submentalis) to the chin; (2) the **inferior labial artery** (a. labialis inferior) to the lower lip; and (3) the **superior labial artery** to the upper lip.

A small vessel, the **transverse facial artery**, crosses the cheek, running along the ventral border of the zygomatic arch. It is a branch of the superficial temporal (p. 242).

(e) The **anterior facial vein** (v. facialis anterior) accompanies the external maxillary artery. It begins in front of the eye as the **angular vein**, and receives as tributaries the **superior** and **inferior labial veins**.

2. Dissection of the facial muscles. These muscles arise from the facial portion of the skull, and are inserted into the skin about the upper and lower lips.

(a) The **quadratus labii superioris**. Origin: Dorsal portion of the maxillary fossa. Insertion: Skin of the upper lip.*

(b) The **subcutaneus faciei**. Origin: Lateral border of the premaxilla, its frontal process, and the supraorbital process of the frontal. Insertion: Skin of the dorsal surface of the nose.

The muscle is closely associated in front of the eye with a conspicuous band of fibres arising from the anterodorsal margin of the zygomatic arch. The latter muscle is probably a corrugator supercilii, operating in conjunction with the orbicularis in firmly closing the eye.

(c) The **zygomaticus minor**. Origin: Anterior end of the zygomatic arch. Insertion: Skin of the angle of the mouth.

(d) The **levator alae nasi**. Origin: Maxillary fossa. Insertion: Lateral cartilage of the nose. The muscle is very slender, and is inserted by a long tendon which underlies the insertion portion of the quadratus labii superioris.

---

*The levator alae nasi, and zygomaticus minor muscles may be considered to be subdivisions of this muscle.

(e) The **buccinator** is a broad stout band of fibres enclosing the cheek and attached to the alveolar borders of the upper jaw and mandible.

(f) The **caninus.** Origin: Lateral border of the upper jaw. Insertion: Hairy portion of the mucous membrane of the mouth. The muscle is very broad, but short and extremely thin.

(g) The **quadratus labii inferioris.** Origin: Ventral border of the mandible. Insertion: Skin of the lower lip.

(h) The **mentalis.** The muscle surrounds the anterior portion of the mandible behind the incisor teeth. It is attached externally to the skin of the lower lip through the insertion portion of the quadratus labii inferioris, which largely overlies it.

Because of the great size and mobility of the ears, the cutaneous auricular muscles, comprising some twenty different members, are especially well developed. These muscles are not individually described, but their development should be noted in contrast to the vestigial character of the ear muscles in man.

3. Dissection on the ventral surface of the neck to free the external jugular vein and its tributaries. The cervical fascia and a portion of the parotid gland must be removed.

The **external jugular vein** (v. jugularis externa) is formed behind the angle of the mandible by the union of the anterior and posterior facial veins. It passes backward in a superficial position to the superior thoracic aperture. Its connections in the lower part of the neck comprise the **transverse scapular vein** (v. transversa scapulae) of the shoulder and its union with the vein of the other side by the **transverse jugular vein** (v. jugularis transversa) (Fig. 82).

The **posterior facial vein** (v. facialis posterior) is formed below the base of the ear by the union of the **superficial** and **deep temporal veins** of the temporal region, the **inferior ophthalmic vein,** from the posterior portion of the orbit, and the **anterior auricular vein** from the ear. It passes downward in the substance of the parotid gland, and in its first portion is covered by the root of the facial nerve. Immediately below the latter it receives the **posterior auricular vein** from the ear and back of the head.

In addition to the tributaries described above, the **anterior facial** vein receives from beneath the anterior margin of the masseter the **deep facial vein** (v. facialis profunda). The latter arises in the lower anterior portion of the orbit, and passes downward beneath the masseter muscle. The anterior facial vein receives at the ventral border of the mandible the **internal maxillary vein** (v. maxillaris interna). The latter also begins in the orbit, where it is connected with the deep facial. At the medial surface of the mandible it receives the **inferior alveolar vein**—to be seen at a later stage—from the interior of the mandible.

The external jugular vein may be divided and turned forward together with the parotid gland.

4. Examination of the more superficial structures of the ventral surface of the head and neck.

(a) The **submaxillary gland** (gl. submaxillaris), one of the salivary series, a somewhat compact rounded or oval gland lying at the medial side of the extreme ventral portion of the angle of the mandible. Its whitish-colored **duct** (d. submaxillaris) may be seen running upward and slightly forward to enter the mouth. It crosses the lateral surface of the digastric muscle but is medial to the external maxillary artery.

(b) The angle of the mandible is covered by two muscles of mastication, the **masseter** lying on the lateral surface, and the **pterygoideus internus** on the medial surface.

(c) The **digastricus.** Only its insertion portion is visible. It passes forward along the medial surface of the mandible, to the anterior portion of which it is attached.

(d) The **mylohyoideus** is a transverse sheet of muscle arising from the medial surface of the mandible on either side and inserted on the hyoid bone.

(e) The **sternomastoideus.** Origin: In common with that of the opposite side, from the manubrium sterni. Insertion: Mastoid process of the skull.

(f) The **sternohyoideus.** Origin: In common with that of the opposite side, from the dorsal surface of the manubrium and anterior portion of the body of the sternum, extending to the third costal articulation. Insertion: Greater cornu of the hyoid.

The two muscles are closely associated in the middle line. They should be separated from one another and divided.

(g) The **sternothyreoideus.** Origin: In common with the sternohyoideus. Insertion: Lateral plate of the thyreoid cartilage of the larynx. The muscle forms a thin band lying on the side of the trachea. It is continued from the thyreoid cartilage to the greater cornu of the hyoid as the **thyreohyoideus.**

(h) The **trachea** occupies a median position, and is supported by cartilaginous **tracheal rings.**

(i) The **thyreoid cartilage** of the larynx; a saddle-shaped cartilage, composed of **right** and **left thyreoid plates** (Fig. 65).

(j) The **cricoid cartilage,** a thick annular cartilage situated between the thyreoid cartilage and the first tracheal ring. It is connected ventrally with the thyreoid cartilage by the cricothyreoideus muscle.

(k) The **deep cervical lymph gland** (lymphoglandula cervicalis profunda) is a large elongated reddish-colored gland in the upper portion of the neck, opposite the thyreoid cartilage.

(l) The **thyreoid gland** (gl. thyreoidea) lies on the ventral surface of the trachea behind the cricoid cartilage. It is composed of right and left portions connected across the middle line by a thin median portion, the **isthmus.**

(m) The **common carotid artery** (a. carotis communis) passes forward from the superior thoracic aperture along the side of the trachea. Its branches on the neck include the **superior thyreoid artery** (a. thyreoidea superior), to the thyreoid gland, and the (superior) **laryngeal artery** (a. larnygea). The latter arises at the level of the thyreoid plate, passing to the larnyx and to the sternohyoid sternothryeoid muscles.

(n) The **internal jugular vein** (v. jugularis interna) lies to the lateral side of the common carotid artery, traversing the neck from the jugular foramen of the skull to the superior thoracic aperture.

(o) The **tenth cranial** or **vagus nerve** (n. vagus) is the largest of four nerves accompanying the carotid artery. It lies to the lateral side of the common carotid, between the latter

and the internal jugular vein. It gives off the **n. laryngeus superior** to the larynx, the latter passing to the dorsal side of the common carotid artery.

(p) The **ramus descendens** of the **twelfth cranial** or **hypoglossal nerve** crosses the root of the vagus from a lateral to a medial position. It passes backward on the ventral surface of the artery, and is chiefly distinguishable by its branches to the sternohyoideus and related muscles.

(q) The **cervical portion** of the **sympathetic trunk** lies on the dorsal surface of the common carotid, and is slightly medial in relation to the vagus.

(r) The **ramus cardiacus** of the **vagus** (n. depressor) lies on the dorsal surface of the common carotid on the medial side of the sympathetic trunk. It arises at the level of the posterior margin of the thyreoid cartilage.

Occurring in the rabbit as a separate nerve, the depressor is important experimentally. Stimulation of the proximal end in the living animal produces fall of blood pressure and retardation of the heart beat. The former is due to a reflex action on the bloodvessels (cf. p. 62), while the latter depends upon reflex stimulation of the vagus, since slowing of the heart does not take place if the vagi are also divided.

(s) The **third** and **fourth cervical nerves** may be traced from their origin in the intervertebral foramina to the musculature of the neck. They encircle the basioclavicularis muscle, under cover of the sternomastoideus and cleidomastoideus.

5. Dissection of the muscles of mastication and related structures of the mandible.

(a) The **masseter muscle.** Origin: The entire surface of the zygomatic arch; tendinous from its anterior angle (spina masseterica). Insertion: Lateral surface of the angle of the mandible (1, a).

The orbital structures should be freed from the zygomatic arch by passing a knife along its dorsal margin. The zygomatic arch may then be divided anteriorly and posteriorly and removed, together with the whole insertion of the masseter muscle.

(b) The **temporalis** is a slender, somewhat triangular muscle arising from the reduced temporal fossa (sulcus temporalis) of the skull and inserted by a long stout tendon on the

medial side of the reduced coronoid process. The muscle may be exposed by dividing the temporal portion of the posterior supraorbital ligament which holds its tendon in place; then divided.

On account of the narrowness of the space lying between the two limbs of the mandible, and the great depth of its angle, it is necessary, in order to expose the surface for the deep dissection of the ventral portion of the head and neck, to remove one-half of the mandible entirely. Hence the following order:

Divide the mandibular symphysis, and pass a knife along the medial surface of the bone to be removed. The tip of the knife must be kept close to the bone, so that the underlying soft parts, except for being divided, will be kept uninjured. The medial surface of the mandible should be clean when removed. The bone may be turned laterad and detached from the skull at the articulation.

The structures appearing on the cut surface include the insertion of the digastricus and the margin of the mylohyoideus; also:

(c) The **pterygoideus internus** muscle. Origin: Pterygoid process of the skull. Insertion: Ventral portion of the medial surface of the angle.

(d) The **pterygoideus externus.** Origin: Lateral plate of the pterygoid process. Insertion: Dorsal portion of the medial surface of the angle. Both muscles are strongly developed.

(e) The **inferior alveolar artery** (a. alveolaris inferior) lies between the two pterygoidei. Its cut end marks the point of entrance into the mandible through the mandibular foramen. The corresponding **inferior alveolar vein** leaves the mandible at this point.

(f) The **inferior alveolar nerve** (n. alveolaris inferior) accompanies the inferior alveolar artery to the mandible. The continuation of the nerve is the **mental nerve.** It appears at the mental foramen, and passes to the lower lip.

The origin of the inferior alveolar nerve may be traced. It arises from the **mandibular nerve** (n. mandibularis), the third division of the **fifth cranial** or **trigeminal nerve** (n. trigeminus), which also gives off anteriorly the stout **lingual nerve** to the tongue and posteriorly the slender

mylohyoid nerve. These structures, together with the inferior alveolar artery may be freed from their loose connections with the pterygoidei, so that they may be left in place for further study. The two pterygoidei may then be detached at their point of origin from the skull and removed.

6. The branches of the common carotid may be traced in the anterior portion of the ventral surface of the neck as follows:

(a) The **internal carotid** (a. carotis interna) is a small vessel given off from the dorsal wall (pp. 133, 278). The trunk then passes forward as the **external carotid** (a. carotis externa).

(b) The **occipital artery** (a. occipitalis) passes from the dorsal wall to the posterior portion of the head.

The **stylohyoideus major**, a slender muscle arising with the digastricus from the **stylohyoid ligament** and inserted on the greater cornu of the hyoid, should be divided. The tendon of the digastricus may be reflected.

(c) The **lingual artery** (a. lingualis) arises from the ventral wall and passes forward into the tongue.

The hypoglossal nerve crosses the ventral surface of the artery and should be kept intact.

(d) The **external maxillary artery** (a. maxillaris externa) is given off immediately in front of the lingual artery, sometimes in common with it. It passes forward on the medial surface of the ventral border of the mandible (medial to the digastricus), giving branches to the submaxillary gland and to the muscles of mastication. The vessel has been divided at the point where it passes around the ventral border of the mandible to the lateral surface of the face.

(e) The **internal maxillary artery** (a. maxillaris interna), one of the two terminal branches of the external carotid, passes in the direction of the orbit (p. 251), giving off the **inferior alveolar artery** to the mandible.

(f) The **superficial temporal artery** (a. temporalis superficialis), the second terminal branch, passes dorsad to the temporal region, supplying the latter and the base of the ear. The **transverse facial artery,** which crosses the cheek, is an anterior branch of this vessel.

7. Dissection of the tongue and hyoid:

The mylohyoideus should be reflected. Note the position of the lingual nerve.

(a) The **stylohyoideus major** muscle. Origin: Jugular process of the occipital bone. Insertion: Tip of the greater cornu of the hyoid. The muscle has been divided.

The superficial temporal and internal maxillary arteries should be divided.

(b) The **styloglossus.** Origin: Jugular process. Insertion: The muscle passes downward and forward, expanding at the base of the tongue into a broad sheet, the fibres of which extend to its anterior tip.

The muscle should be carefully separated from two others on its dorsomedial side and divided.

(c) The **stylohyoideus minor.** Origin: Jugular process. Insertion: Lesser cornu of the hyoid. A slender muscle having about the same direction, but ending on the more dorsal part of the hyoid apparatus.

The remaining muscle is the **stylopharyngeus**, a thin delicate muscle, the insertion of which is on the lateral wall of the pharynx.

(d) The **geniohyoideus.** Origin: Mandibular symphysis. Insertion: Anterior surface of the body of the hyoid: unpaired.

(e) The **genioglossus.** Origin: Medial surface of the mandible immediately behind the symphysis. The fibres pass upward and slightly backward into the substance of the tongue.

(f) The **hyoglossus.** Origin: The body of the hyoid and the greater and lesser cornua by more or less separate heads. The muscle passes into the base of the tongue, enclosed on either side by the styloglossi.

(g) The **lingualis,** or intrinsic muscle of the tongue consists of a mass of fibres with no skeletal attachments.

(h) The **lingual nerve** (n. lingualis), one of the chief branches of the **mandibular,** passes forward and downward to the side of the tongue and enters its substance immediately below the ventral border of the styloglossus.

The lingual is the sensory nerve of the tongue. Near its point of origin the lingual nerve receives the chorda tympani (p. 254).

(i) The **twelfth cranial** or **hypoglossal nerve** (n. hypoglossus) enters the base of the tongue. It lies on the lateral side of the external carotid artery and on the medial side of the stylohyoideus major. It is distributed as a motor nerve to the lingual muscles.

(j) The **ramus lingualis** of the **ninth cranial,** or **glossopharyngeal** nerve (n. glossopharyngeus) enters the base of the tongue at a point dorsal to the hypoglossus and between the stylohyoideus minor and the stylopharyngeus. It is a gustatory nerve of the tongue.

8. Dissection of the extra-cranial roots of the ninth to twelfth nerves (Fig. 79).

These nerves, which for the most part have already been exposed, may be traced to their origin in the jugular and hypoglossal foramina. The tympanic bulla should be cleared and the tendons of origin of tongue muscles removed from the jugular process.

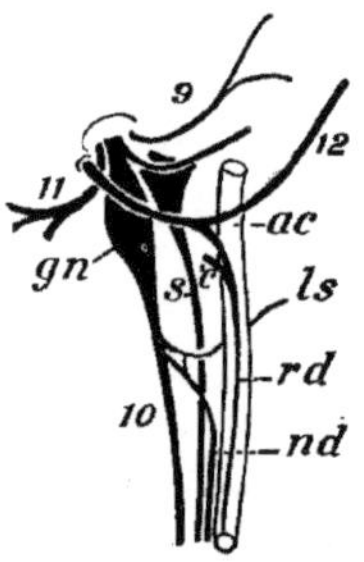

FIG. 79. Plan of the extra-cranial roots of the IX-XII cranial nerves and sympathetic trunk; ventral surface, right side, the sympathetic and depressor nerve shown as displaced from the dorsal surface of the artery. 9, 10, 11, 12, glossopharyngeal, vagus, spinal accessory, and hypoglossal nerves; ac, carotid artery; c, cervical root of ramus descendens XII; gn, ganglion nodosum vagi; ls, superior laryngeal; nd, depressor nerve; rd, ramus descendens hypoglossi; s, sympathetic.

(a) The **ninth** (glossopharyngeal) nerve is farthest forward. Its two main branches are the **ramus lingualis** to the tongue, and the **ramus pharyngeus,** the latter entering the lateral wall of the pharynx.

(b) The **tenth** (vagus) nerve bears an elongated ganglionic enlargement, the **plexus ganglioformis** (ganglion nodosum). It lies immediately below the jugular foramen.

The **superior laryngeal nerve** and the **ramus cardiacus** (depressor nerve) are given off at the level of the origin of the internal carotid artery.

(c) The **eleventh cranial,** or **spinal accessory nerve** (n. accessorius) is dorsal to the vagus. The nerve passes dorsad to the medial side of the mastoid attachments of the sternohyoideus and cleidomastoideus muscles, giving branches to the latter, and then passes backward to the ventral surface of the trapezius to which it is distributed.

(d) The **twelfth** (hypoglossal) nerve arises behind the foregoing nerves, since it comes from the hypoglossal foramina of the occipital. It crosses their roots, forming a broad curve on the lateral surface of the root of the external carotid artery and enters the base of the tongue. The **ramus descendens** is given off at about the point where it crosses the artery. It has a slender root from the third cervical nerve.

(e) The **cervical portion** of the **sympathetic trunk** begins in the **superior cervical ganglion** (g. cervicale superius). It lies to the medial side of the vagus ganglion and of the internal carotid artery. The nerves proceeding from the ganglion accompany the branches of the external and internal carotid arteries to the head.

9. The oral cavity and pharynx.

The glossopharyngeal nerve and the superior laryngeal nerve and artery may be divided, and the external carotid artery with the associated nerves separated from the oesophagus and trachea. The latter may be displaced downward to a slight extent by dividing the loose connective tissue along the ventral surface of the vertebral column. An incision extending from the oral cavity backward into the oesophagus will expose the internal surface of this portion of the digestive tube sufficiently to make out its features. The incision divides the **constrictor pharyngis muscle,** a broad band of muscle fibres enclosing the posterior portion of the pharynx.

For the general relations of the oral cavity see p. 80 and Fig. 41.

The divisions are:

(a) The **oral cavity** (cavum oris), divisible into the **oral cavity proper,** and the **vestibulum oris,** the latter laying between the alveolar processes and teeth on the one hand and the lips on the other.

(b) The **pharynx:** its **oral portion** (pars oralis) continues the oral cavity, and connects it with the oesophagus. Its dorsal, also anterior, **nasal portion** (pars nasalis), or **nasopharynx,** lies above the soft palate, and receives the posterior aperture of the nose. Its ventral and posterior **laryngeal portion** (pars laryngea), not well-defined, contains the aperture of the larynx, the **aditus laryngis.**

In the oral cavity:

(a) The **hard palate** (palatum durum) forms the anterior portion of the roof; its mucous membrane is thrown into a series of transverse ridges.

(b) The **soft palate** (palatum molle) is the thin, narrow, posterior, membranous portion of the roof. It is very long in the rabbit, extending from the bony palatine bridge backward to a point above the laryngeal aperture, where it ends with a concave free margin.

(c) The **nasopalatine** or **incisive ducts** (dd. nasopalatini) open by minute apertures immediately behind the small incisors. They connect the anterior portion of the nasal cavity with the mouth.

(d) The **tongue** (lingua) projects upward and forward from its basal attachments on the hyoid into the floor of the mouth. Its connection with the latter is extended forward in the middle line by a vertical membranous fold, the **frenulum linguae.** Its dorsal surface is divided into a posterior smooth and hard portion, and an anterior softer and rougher portion, occupied by fine low elevations, the **fungiform papillae** (papillae fungiformes). At the posterior end of the smooth portion there is on either side a minute spherical elevation, set low into the mucous membrane, the **vallate papilla** (papilla vallata), and in a more lateral and anterior position an oval area, the **papilla foliata,** the surface of which is marked by fine parallel ridges. Both vallate and foliate papillae are occupied by microscopic taste-buds.

In the pharynx:

(a) The **tonsil** (tonsilla) appears as a rounded mass of lymph follicles lying on the anterior wall of a deep lateral depression, the **tonsillar sinus** (sinus tonsillaris). The vertical slit-like aperture of the sinus is bounded by low anterior and posterior folds.

(b) The **epiglottis,** a valve-like fold guarding the entrance to the larynx, projects upward from the floor into the pharyngeal cavity.

(c) By removing the posterior portion of the soft palate, the connection of the nasopharynx with the nasal fossae will be

exposed; also on the lateral wall the **pharyngeal aperture** of the **auditory tube** (ostium pharyngeum tubae).

10. Examination of the larynx.

By cutting around the base of the tongue on the opposite side of the body, the whole structure, together with the hyoid, larynx, and a portion of the trachea back to about the end of the thyreoid gland may be removed. This affords a good opportunity of re-dissecting on the opposite side from the medial surface of the mandible outward, also of clearing and examining the hyoid apparatus, which is not usually available with the prepared skeleton (Fig. 65). The laryngeal cartilages should be cleared externally and the parts made out as follows:

(a) The **thyreoid cartilage** (cartilago thyreoidea) forms the largest portion of the structure. It is an unpaired saddle-shaped cartilage, described as consisting of **right** and **left laminae.** Its anterodorsal angle projects forward as the **cornu superior**; connected by ligament with the greater cornu of the hyoid. The corresponding posterodorsal angle, the **cornu inferior,** overlies the dorsolateral portion of the cricoid cartilage. The anterior dorsal portion of each plate bears a small **thyreoid foramen** (foramen thyreoideum) for the entrance of the superior laryngeal nerve.

(b) The **cricoid cartilage** (cartilago cricoidea) is an annular cartilage, partly enclosed by the posterior portions of the thyreoid laminae and surrounding the first tracheal ring. Its ventral portion, the **arch** of the cricoid cartilage, falls some distance behind the thyreoid cartilage, the intervening space being largely occupied by the **cricothyreoidei muscles.** Its dorsal portion, the **lamina** of the cricoid, forms a greatly expanded plate serving for the support of the arytenoid cartilages.

(c) The paired **arytenoid cartilages** (cartilagines arytenoideae) lie one on either side of the anterior tip of the cricoid plate.

(d) The **corniculate cartilages** (cartilagines corniculatae) are slender curved terminal cartilages supported by the arytenoids.

(e) The **epiglottic cartilage** (cartilago epiglottica) is a triangular plate supporting the epiglottis.

(f) The rudimentary **vocal folds** (plicae vocales) may be seen as vertical folds of the internal surface of the larynx, especially prominent when the thyreoid cartilage is bent downward on thö cricoid. Each fold forms the posterior boundary of a shallow pouch, the **laryngeal ventricle** (ventriculus laryngis).

In addition to the criothryreoidei, the laryngeal cartilages are connected by several small muscles, including the cricoarytenoidei posterior and lateralis, the thryreoarytenoideus and the arytenoideus transversus, the last named being an unpaired muscle connecting the arytenoid cartilages.

11. The eye and related structures of the orbital cavity.

The eyeball should be carefully separated from the bony orbital rim. The first portion of the nasolacrimal duct (d. nasolacrimalis), passing from its aperture on the medial surface of the lower eyelid to the lacrimal bone, will be divided. The muscles and glands of the orbit may be made out as follows:

(a) The **levator palpebrae superioris.** Origin: Wall of the orbit above the optic foramen. Insertion: Upper eyelid.

This thin sheet of muscle should be separated from the underlying rectus superior of the eyeball.

(b) The **obliquus superior.** Origin: Anterior margin of the optic foramen. The muscle passes upward on the wall of the orbit, then beneath a fibrous cord, the trochlea, which bridges a small portion of the orbital wall and changes the course of the tendon by a considerable angle. Insertion: Anterodorsal portion of the eyeball.

(c) The **obliquus inferior.** Origin: Lacrimal bone. Insertion: Posteroventral portion of the eyeball.

(d) The four **recti muscles, superior, inferior, medialis,** and **lateralis,** arise from the boundary of the optic foramen, and are inserted respectively on the dorsal, ventral, anterior, and posterior portions of the periphery of the eyeball.

(e) The **retractor oculi** (best seen after the removal of the eye) arises in common with the foregoing recti muscles, and is

inserted on the medial portion of the eyeball around the optic nerve. It consists of four distinct parts.

(f) The **Harderian gland** (gl. Harderiana) is a large compact gland lying in the anterior portion of the orbit. It is composed of two portions, one of which is greyish red, the other white. The duct opens on the inner surface of the third eyelid.

(g) The **lacrimal gland** (gl. lacrimalis) is a much smaller, also darker, structure lying close to the skull in the temporal angle of the orbit. It communicates by several fine ducts with the inner surface of the upper eyelid.

The **infraorbital gland** (gl. infraorbitalis) is a diffuse white or yellowish gland lying in the anteroventral angle of the orbit immediately medial to the zygomatic arch. The gland is one of the salivary series, its duct passing downward and opening through the mucous membrane of the cheek into the cavity of the mouth.

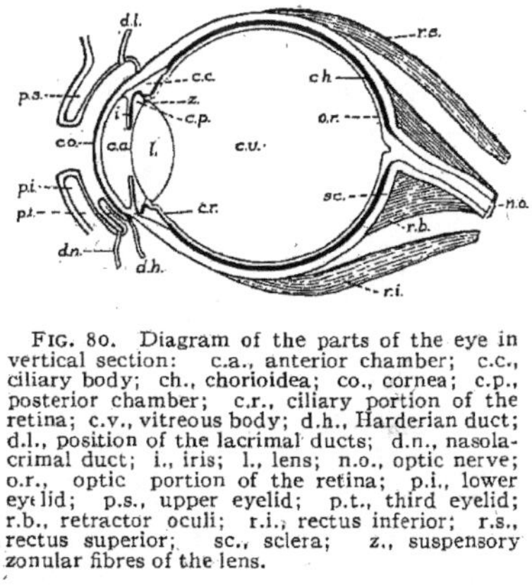

FIG. 80. Diagram of the parts of the eye in vertical section: c.a., anterior chamber; c.c., ciliary body; ch., chorioidea; co., cornea; c.p., posterior chamber; c.r., ciliary portion of the retina; c.v., vitreous body; d.h., Harderian duct; d.l., position of the lacrimal ducts; d.n., nasolacrimal duct; i., iris; l., lens; n.o., optic nerve; o.r., optic portion of the retina; p.i., lower eyelid; p.s., upper eyelid; p.t., third eyelid; r.b., retractor oculi; r.i., rectus inferior; r.s., rectus superior; sc., sclera; z., suspensory zonular fibres of the lens.

To examine the structure of the eye, the muscles of the eyeball should be divided at their insertions, and the whole structure removed. The **second cranial** or **optic nerve** (n. opticus) is divided; also the **ophthalmic artery**, a small branch of the internal carotid which accompanies the nerve outward from the optic foramen to the eyeball.

The eye may be divided by a circular incision into medial and lateral hemispheres, the lateral hemisphere, containing the lens, being again divided vertically. The parts should be examined under water. The chief structures (Fig. 80) comprise:

(a) The **fibrous tunic** (tunica fibrosa oculi), the strong peripheral coat enclosing the whole structure. It is divisible into a medial portion, the **sclera**, or **sclerotic coat**, a thick

white investment of fibrous connective tissue enclosing the greater part of the eyeball, and a smaller transparent lateral portion, the **cornea**, covering the exposed surface.

(b) The **vascular tunic** (tunica vasculosa oculi), the thin middle coat of the eye; pigmented, except in albino animals. It is divisible into: (1) a general portion, the **chorioidea**, lying on the inner surface of the sclera; (2) a muscular portion, the **ciliary body** (corpus ciliare), forming an annular ridge about the periphery of the lens; and (3) the **iris**, the latter forming a circular fold suspended about the periphery of the lens and on its outer surface.

(c) The **retina**, the innermost layer of the eye, forms a thin soft membrane covering the inner surface of the chorioidea. It is divisible in a larger **optic portion**, the sensory part of the eye, and a smaller **ciliary portion**, lying about the periphery of the lens and distinguishable by the radiate markings of its surface, the latter formed by the projecting ridges of the ciliary body.

(d) The transparent **lens** of the eye is suspended by fine filaments, the **zonular fibres**, reflected from the margin of the ciliary body.

(e) The **vitreous body** (corpus vitreum), a transparent mass, of gelatinous consistence, occupies the large space enclosed by the lens and the retina.

(f) The space enclosed between the surface of the lens and the cornea is divisible into a larger portion, the **anterior chamber** (camera oculi anterior), lying outside of the iris, and a smaller portion, the **posterior chamber** (camera oculi posterior), lying between the iris and the lens. The central space enclosed by the free margin of the iris is the **pupil** (pupilla).

12. Following the removal of the eye, the bloodvessels and nerves of the orbit may be freed from the remaining portions of the eye muscles and examined. In order to see their connections in the anterior angle of the orbit, it is necessary to break away the anterior root of the zygomatic arch, and also the bony ridge which lodges the alveoli of the posterior cheek-teeth.

(a) The **internal maxillary artery** enters the orbit through the anterior sphenoidal foramen in the root of the lateral lamina of the pterygoid process. At the posterior ventral angle of the orbit it gives off the **inferior ophthalmic artery** (a. ophthalmica inferior). This vessel passes upward and forward on the medial wall of the orbit, giving branches to the eye muscles. It divides into two branches, the **frontal artery,** which leaves the orbit through the anterior foramen of the supraorbital process, and the **lacrimal artery,** which passes through the corresponding posterior foramen. The ethmoidal artery, a small branch of the frontal, passes through the minute ethmoidal foramen of the orbital portion of the frontal into the nasal cavity.

The internal maxillary artery passes forward along the ventral boundary of the orbit, and at the opening of the infraorbital canal gives off a branch, the pterygopalatine artery, continuing as the infraorbital artery. A small branch, the **superior dental artery** (a. dentalis superior) is given off laterally to the alveoli of the upper teeth.

The **infraorbital artery** (a. infraorbitalis) passes through the infraorbital canal to the face.

The **pterygopalatine artery** (a. pterygopalatina) divides almost immediately into the **anterior palatine artery,** which traverses the pterygopalatine canal to the mucous membrane of the hard palate, and the **sphenopalatine artery,** which enters the nasal cavity by the sphenopalatine foramen.

(b) The divisions of the **third cranial,** or **oculomotor nerve,** supply the eye muscles, with the exception of the obliquus superior, rectus lateralis, and retractor oculi.

This nerve enters the orbit from the superior orbital fissure in company with certain parts of the trigeminal (e,f). The small nerves passing through the middle and posterior sphenoidal formina of the pterygoid process are the **pterygobuccinator** and **masseterico-temporal** nerves, branches of the mandibular, passing to the muscles of mastication.

(c) The **fourth cranial,** or **trochlear nerve** (n. trochlearis), is distributed to the obliquus superior muscle.

(d) The **sixth cranial**, or **abducent nerve** (n. abducens), is distributed to the rectus lateralis, and to the retractor oculi.

(e) The **ophthalmic nerve** (n. ophthalmicus), the first division of the **fifth cranial**, or **trigeminal nerve** (n. trigeminus), accompanies the inferior ophthalmic artery on the medial wall of the orbit. It gives off a **lacrimal nerve**, which passes upward through the posterior foramen of the supraorbital process, and then passing forward a short distance divides in two parts. One of these, the **frontal nerve**, leaves the orbit through the anterior supraorbital foramen, while the other, the **nasociliary nerve**, is partly distributed to the anterior portion of the orbit, and is connected with the minute ciliary ganglion lying on the optic nerve, forming its sensory root. Its chief portion leaves the orbit through the ethmoidal foramen of the orbital portion of the frontal bone as the **ethmoidal nerve**.

The lacrimal, frontal and nasociliary nerves appear as separate structures in the orbit, their origin being deep.

(f) The branches of the **maxillary nerve** (n. maxillaris), the second division of the trigeminus, traverse the ventral portion of the orbit passing forward in company with the internal maxillary artery. They include the **sphenopalatine nerve** (n. sphenopalatinus) and the **infraorbital nerve** (n. infraorbitalis). The latter gives off superior alveolar branches to the upper teeth, passing forward through the infraorbital groove and foramen to the face.

The connections of the sphenopalatine nerve may be examined by first dividing both nerves at the posterior angle of the orbit; then separating the slender sphenopalatine nerve from the ventral surface of the cord and turning the principal, infraorbital portion forward. A third nerve, the nerve of the pterygoid canal, should remain intact on the orbital wall. If the infraorbital nerve alone is divided, the sphenopalatine nerve will be found on the surface of the bone below the nerve of the pterygoid canal, from which it may be distinguished by its lighter coloration.

The sphenopalatine nerve is continued forward as the **anterior** (major) **palatine nerve**, which passes through the pterygopalatine canal to the posterior portion of the hard palate, but is also con-

nected with the **sphenopalatine ganglion.** **Nasal rami** pass to the mucous membrane of the nose, and the **nasopalatine nerve** enters the nasal region, traversing the surface of the septum and reaching the anterior portion of the palate through the incisive foramina.

The **nerve** of the **pterygoid canal** (n. canalis pterygoidei) is a slender cord which passes backward along the orbital wall from the posterodorsal angle of the ganglion. It lies on the medial side of the sphenopalatine and infraorbital nerves and on the lateral surface of the palatine bone. Posteriorly it enters the groove representing the pterygoid canal.

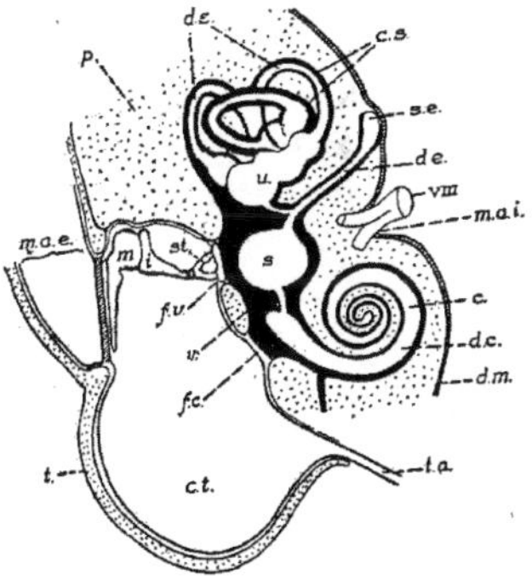

FIG. 81. Diagram of the parts of the ear in vertical projection. To show the general relations of the structures covered by the dissection.

p, petrous portion of the petrotympanic bone; t., tympanic portion (bulla tympani).

c., cochlea; c.s., bony semicircular canals; c.t., tympanic cavity; d.c., cochlear duct; d.e., endolymphatic duct; d.m., dura mater; d.s., semicircular ducts; f.c., cochlear fenestra; f.v., vestibular fenestra; i., incus; m, malleus; m.a.e., external acoustic meatus, terminating at the tympanic membrane; m.a.i., internal acoustic meatus; s., sacculus; s.e., endolymphatic sac; st., stapes; t.a., auditory tube; u., utriculus; v, vestibulum; VIII, acoustic nerve.

This nerve is composed of two parts, separated posteriorly. One of them, the **deep petrosal nerve** (n. petrosus profundus), is connected with the sympathetic plexus of the internal carotid artery. The other, the **great superficial petrosal nerve** (n. petrosus superficialis major) enters the skull at the foramen lacerum, passing into the petrosal bone, in the interior of which it is connected with the trunk of the facial. The nerve constitutes the motor root of the sphenopalatine ganglion, the sensory root being that provided by the sphenopalatine nerve.

The sphenopalatine ganglion is one of several representing the sympathetic system in the head, and having motor and sensory roots from the cerebral nerves in addition to their sympathetic connections. The series includes the **ciliary ganglion,** which lies on the optic nerve, the **sphenopalatine ganglion,** the **otic ganglia,** associated with the mandibular

nerve, and the **submaxillary ganglion,** associated with the lingual nerve.

13. Examination of the middle ear.

By breaking away the ventrolateral portion of the tympanic bulla and clearing the surface, the structures of the tympanic cavity may be studied. They are chiefly those already described in connection with the skeleton (p. 147), but the following soft parts may be identified.

(a) The **tympanic membrane** (membrana tympani) is stretched almost vertically across the lower end of the external acoustic meatus.

(b) The **tensor tympani** is a slender muscle, the origin of which forwards from the alisphenoid is concealed. It is inserted on the manubrium mallei.

(c) The **stapedius** is a minute muscle arising from the periotic bone above the cochlear fenestra and inserted on the stapes.

(d) The **chorda tympani** is a delicate nerve which crosses the tympanic cavity, lying between the long crus of the incus and the manubrium mallei.

The nerve is a continuation of the **intermediate nerve,** a sensory root of the facial, which arises independently of the chief or motor root and joins the facial in the facial canal of the periotic bone. Its peripheral connections are with the lingual nerve and the submaxillary ganglion.

(e) The **internal carotid artery** traverses the carotid canal of the tympanic bone. By breaking away the posterior portion of the bulla, the entrance of the vessel into the external carotid foramen may be seen.

The dissection of the parts of the ear as here outlined includes only the external and middle portions together with the associated acoustic nerve and its entrance to the periotic bone. The parts of the **internal ear** (Fig. 81) are not readily made out without the use of special methods, though their position may be estimated by carefully breaking away the surface of the ventral portion of the periotic. They include (1) the bony **labyrinth,** consisting of a series of connected spaces lodged in the interior of the petrous bone, and comprising the **cochlea, vestibulum,** and the **bony semicircular canals**; and (2) the **membranous labyrinth,** consisting of a second series of spaces contained within the first, and comprising the **duct of the cochlea,** the **sacculus,** the **utriculus,** and the **semicircular ducts,** together with their connections and the **endolymphatic duct** and **sac.** The membranous labyrinth contains the sensory portion of the ear and its cavity is occupied by a fluid material, the **endolymph.** The wall is separated from that of the bony labyrinth by an extensive **perilymphatic space** also occupied by a fluid material termed the **perilymph.**

XI. THE THORAX.

1. Examination of the thoracic wall. For this purpose the lateral surface of the thorax may be conveniently cleared, on the side from which the anterior limb has been removed, by dissecting away the attachments of the muscles already examined in the previous dissections. These include the origins of the pectorales, pecto-scapularis, serratus anterior (thoracic portion), obliquus externus, and rectus abdominis.

The dorsal portion of the exposed surface is occupied by the spinal musculature; to be examined at a later stage. On the ventral portion appear the ribs, and between them, filling the intercostal spaces, the intercostal muscles. The **external intercostals** (mm. intercostales externi) arise from the posterior margins of the bone ribs, the fibres passing obliquely downward and backward to be inserted on the anterior margins of the next succeeding ribs. The **internal intercostals** (mm. intercostales interni), the fibres of which are disposed in the opposite direction, are concealed for the most part from this surface, but appear ventrally between the costal cartilages, where they are not covered by the external intercostals. They are best examined at a later stage from the interior of the thorax.

In preparation for the removal of a section of the thoracic wall, the pectorales should be divided on the opposite side of the thorax, close to the sternum, so that the limb may be displaced.

The nerves and vessels of the neck must be kept intact until the following dissection accounts for their thoracic connections.

The scaleni muscles (p. 268) should be examined, since it is necessary to destroy their costal insertions.

A triangular section of the wall, including the sternum and the costal cartilages, may be removed by making three incisions, one on either side of the sternum, extending from the middle of the first rib backward to the end of the ninth bone rib, and the third across the ventral surface close in front of the diaphragm.

The **transversus thoracis** muscle appears on the inner surface of the section removed. It is a thin sheet of fibres arising from the body and xiphoid process of the sternum and inserted on the ribs,

from the second to the sixth, at the junctions of the bone ribs with the costal cartilages.

The artery passing along the ventral wall between the foregoing muscle and the internal intercostals is the **internal mammary** (p. 257).

For the general relations of heart and lungs, see pp. 84 and 87.

2. Dissection of structures in the superior thoracic aperture.

The nerves and bloodvessels of this region are concealed by the **thymus gland,** a large triangular flattened structure of fatty consistence, extending forward from the base of the heart to the anterior end of the thorax. The thymus should be carefully scraped away, all vessels except those of the gland itself being kept intact.

The following structures, including the aortic arch and the arteries arising from it, the superior caval veins, and the vagus, phrenic, and sympathetic nerves, cannot be dissected exactly in the order given below, but must be separated from one another and identified as they appear. The left superior caval vein is superficial, crossing the ventral surface of the aortic arch. Care should be taken not to injure the nerves (c-f) in exposing the branches of the subclavian artery.

(a) The **arch of the aorta** (arcus aortae). Beginning at the base of the heart, the aorta passes at first forward, and then describing a curve, in the course of which it lies slightly to left of the median plane, turns backward along the ventral surfaces of the bodies of the thoracic vertebrae. With the exception of the coronary arteries (p. 260) the first branches are the large paired vessels arising from the anterior wall. They comprise the **common carotid** and **subclavian arteries.** On the right side the carotid and subclavian arise from a short common trunk, the **innominate artery** (a. anonyma). The left common carotid arises immediately to the left of this vessel or from its base.

The **subclavian artery** (a. subclavia) is the first portion of the artery of the anterior limb. It passes from its point of origin laterad to the anterior margin of the first rib, where it is replaced by the axillary artery. Near its point of origin it gives off several branches, the relations of which are subject to considerable variation. They include:

(1) The **vertebral artery** (a. vertebralis). This vessel passes into the costotransverse foramen of the sixth cervical vertebra, and traversing the canal formed by this and the corresponding foramina of the remaining cervical vertebrae, reaches the interior of the cranial cavity. Its union on the ventral surface of the medulla oblongata with its fellow of the opposite side to form the basilar artery will be seen at a later stage (p. 278).

(2) The **superficial cervical artery** (a. cervicalis superficialis) —divided in a previous dissection (p. 206)—is a small vessel which passes forward and outward beneath the insertions of the cleidomastoideus, basioclavicularis and levator scapulae major muscles, ramifying extensively in the fat mass of the side of the neck under cover of the superior portion of the trapezius. Its **ascending cervical** branch lies on the lateral side of the external jugular vein.

(3) The **transverse artery of the neck** (a. transversa colli), also divided in a previous dissection (p. 207), passes laterad around the neck of the first rib to the wall of the thorax. It passes through the loop formed by the eighth cervical and first thoracic spinal nerves. It runs dorsad, first on the medial side of the scalenus anterior, then on the medial side of the cervical portion of the serratus anterior. A strong branch passes to the inferior angle of the scapula. The artery supplies the serratus anterior and the rhomboidei.

(4) The **a. intercostalis suprema** passes backward to the internal surface of the thoracic wall, giving off the first three (or four) **intercostal arteries** in the intercostal spaces, and also small branches to the oesophagus and trachea.

(5) The **internal mammary artery** (a. mammaria interna), the first portion of which has been removed with the ventral wall of the thorax, passes backward to the ventral abdominal wall as the **superior epigastric artery** (a. epigastrica superior) anastomosing with the inferior epigastric (p. 203).

(b) The **superior caval vein** (v. cava superior) is formed at the base of the neck by the union of the internal and external jugular veins, the latter vessel receiving at this point the **subclavian vein** (v. subclavia). The right superior caval

passes almost directly backward, crossing the ventral surface of the right subclavian artery, and enters the anterior portion of the right atrium. The left vessel crosses both the left subclavian artery and the arch of the aorta, reaching the right atrium from the dorsal surface of the heart.

(c) The **vagus nerve.** On the right side the nerve crosses the ventral surface of the subclavian artery, passing dorsad to the wall of the oesophagus. It gives off the **recurrent nerve** (n. recurrens), the latter curving around the subclavian artery and passing forward along the side of the trachea to the larynx. On the left side the vagus passes between the arch of the aorta and the base of the heart to the ventrolateral wall of the oesophagus. The recurrent nerve passes forward on the dorsal side of the arch.

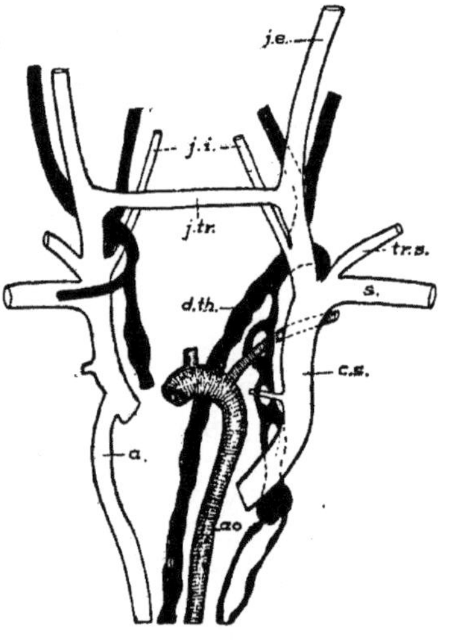

FIG. 82. Plan of the venous and lymphatic trunks of the anterior portion of the body. After McClure and Silvester.
a., azygos vein; ao., aorta; c.s., left superior caval vein; d.th., thoracic duct; j.e., j.i., and j.tr., external, internal and transverse jugular veins; s., left subclavian vein; tr.s., transverse scapular vein.

(d) The **ramus cardiacus** of the vagus. In front of the subclavian artery the nerve is at first closely associated with the vagus trunk, lying on its medial side. On the right side it passes to the dorsal surface of the subclavian, and on the left to the dorsal surface of the aortic arch. It ends in the **cardiac plexus** (plexus cardiacus), a network of sympathetic nerves lying between the aortic arch and the pulmonary artery.

(e) The **phrenic nerve** (n. phrenicus) is a stout cord arising chiefly from the fourth cervical spinal nerve. That of the

left side crosses the ventral surface of the subclavian artery and the aortic arch, passing then along the pericardium to the diaphragm. That of the right side accompanies the thoracic portion of the inferior caval vein. The nerve controls the respiratory movements of the diaphragm.

(f) The **sympathetic trunk.** At the base of the neck the cervical portion of the sympathetic trunk enters the **inferior cervical ganglion** (g. cervicale inferius). The latter lies in front of and somewhat dorsal to the subclavian artery. The **first thoracic ganglion** lies behind the artery and is connected with the inferior cervical by the **ansa subclavia,** a loop formed by two cords, one of which passes to the dorsal, the other to the ventral side of the subclavian artery.
The nerves proceeding from the inferior cervical ganglion enter the cardiac plexus and the sympathetic plexuses of the subclavian and its branches.

3. Dissection of the heart.

The character and relations of the enclosing serous sac, the **pericardium,** should first be noted. Its relation to the heart is similar to that of the peritoneum and pleura investing other visceral organs (p. 99). It comprises a **parietal layer,** that portion commonly known as the pericardium, and a **visceral layer,** the **epicardium,** which forms an immediate investment for the heart substance. The parietal layer forms a loose, capacious sac, the serous membrane being greatly strengthened by the presence of a thin layer of connective tissue which forms an external investment and is usually considered as part of the pericardium.

The paired pleural cavities containing the lungs are broadly separated by a longitudinal vertical partition, the **mediastinum** or **mediastinal septum,** the space enclosed by the latter being largely occupied by the heart and by the cavity of the pericardium. For a considerable area ventrally the pericardium is loosely applied to the thoracic wall, the intervening space, which is bounded laterally by the membrane lining the pleural cavities, being known as the **anterior mediastinum.** A corresponding dorsal space lying between the heart and the bodies of the thoracic vertebrae, and also bounded laterally by the pleura, is the **posterior mediastinum.** It is occupied by several structures, namely, the oesophagus, the thoracic aorta, the bronchi, and the pulmonary bloodvessels.

The pericardium should be removed, and the external features of the heart and its great vessels examined as follows:

(a) The posterior, somewhat conical, **ventricular portion** of the heart. The **left ventricle** (ventriculus sinister) may be distinguished both by its position and by the more solid character of its wall. The **right ventricle** (ventriculus dexter) is less muscular, and the wall is readily pressed inward. The line of division is indicated on the ventral surface by a faint depression, the **anterior longitudinal sulcus.**

(b) The **pulmonary artery** (a. pulmonalis) leaves the base of the right ventricle, passing forward and to the left in a somewhat spiral fashion around the aorta. On the dorsal surface of the latter it divides into the **right** and **left pulmonary arteries**, one for each lung. At the point of division the pulmonary artery is connected with the aorta by a short fibrous cord, the **arterial ligament** (lig. arteriosum), representing the foetal connection of the two vessels through the ductus arteriosus (p. 89).

(c) The **left coronary artery** (a. coronaria sinistra) passes backward on the ventral surface of the heart, lying in the anterior longitudinal sulcus. A corresponding **right coronary artery** (a. coronaria dextra) passes to the right side of the heart, lying between the right ventricle and the right atrium.

The two vessels supply the substance of the heart. They are the first branches of the aorta, arising from the aortic sinuses at its base.

(d) The **left atrium** (atrium sinistrum) is the thin-walled chamber lying to the left at the base of the heart. The **pulmonary veins** (venae pulmonales), several on either side, enter the left atrium, passing from the medial portions of the lungs.

(e) The **right atrium** (atrium dextrum) resembles the left in the character of its wall. It receives the **right** and **left superior caval veins** and the unpaired **inferior caval vein.**

The heart may be removed by dividing the great bloodvessels. The arch of the aorta should be removed with the heart by dividing the vessel at a point beyond the origin of the left subclavian, and then severing the carotids and subclavians at their bases. This exposes the surface for the subsequent examination of the posterior end of the trachea and its connections with the lungs.

Open the right ventricle by a longitudinal incision of the ventral wall, extending the incision forward into the pulmonary artery, also both atria by transverse incisions. By washing out the cavities, the internal features of the wall, including the arrangement of the valvular structures, may be examined as follows:

In the right ventricle:

(a) The **trabeculae carneae;** muscular ridges of the internal surface of the wall.

(b) The **tricuspid valve** (valvula tricuspidalis). The thin membranous flaps composing the valve enclose the atrioventricular aperture, and project into the cavity of the ventricle. Their margins, which are otherwise free, are connected by slender fibrous cords, the **chordae tendineae,** with the **papillary muscles** (mm. papillares), the latter being thick muscular projections, of somewhat conical shape, arising from the opposite walls.

In the rabbit the valve is composed of only two flaps, of which the ventral one is very free, and has large papillary muscles, while the dorsal one is closely attached to the wall, with the papillary muscle reduced or absent. For this reason the term **right atrioventricular valve** is more appropriate than "tricuspid".

(c) The **semilunar valves** (valvulae semilunares) of the pulmonary artery are three extremely thin folds guarding the entrance to the vessel from the right ventricle. Two of the valves are usually found intact, the third being destroyed on opening the vessel.

In the atria:

(a) The respective positions of the pulmonary and systemic veins at their points of entrance.

(b) The complete separation of the two chambers. In the partition separating them may be seen a thin fibrous portion denoting the position of the embryonic **foramen ovale.**

Open the left ventricle by a ventral longitudinal incision, cutting well through the tip of the ventricle and extending the incision across the pulmonary artery and into the aorta. On account of the great thickness of the wall the internal structure is not so easily examined as in the right ventricle.

(a) The **bicuspid** or **left atrioventricular valve** (valvula bicuspidalis) is similar in general structure to the tricuspid valve of the right ventricle, but is more nearly circular in form, with stout, closely grouped papillary muscles.

(b) The **semilunar valves** of the aorta are three in number, and are similar to those of the pulmonary artery.

4. Examination of the lungs and their connections (Figs. 44, 83).

The removal of the ventral wall of the thorax opens the pleural cavities by taking away a considerable portion of the costal pleura, which is adherent to the internal surfaces of the ribs. The chief features may be made out as follows:

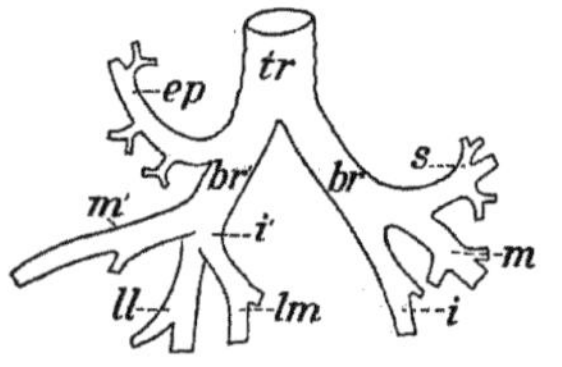

FIG. 83. Plan of the respiratory tubes as seen from the ventral surface. tr, trachea; br, br', left and right bronchi; ep, eparterial bronchus; s,m, m', i, i', bronchial rami to superior, middle and inferior lobes; ll, lm, bronchial rami to lateral and medial lobules.

(a) Each **pleural cavity** (cavum pleurae) is a closed serous sac, the lining membrane, or **pleura**, being distributed over the costal surface as the **costal pleura**, partly over the anterior surface of the diaphragm as the **diaphragmatic pleura**, and over the surface of the lung as the **pulmonary pleura**. Posteriorly the pulmonary pleura passes from the medial margin of the left lung to the medial lobule of the right and thence backward to the diaphragm, forming a broad sheet of attachment, the **pulmonary ligament** (lig. pulmonale).

(b) The **lungs** (pulmones) are paired expansible structures, the surfaces of which are free, except posteriorly, where they are attached to the diaphragm through the pulmonary ligament, and medially where they are connected with the pulmonary bloodvessels and the respiratory tubes.

(c) Each lung is divided into three portions, the **superior**, **middle**, and **inferior lobes**. On account of the smaller size of the left lung, the superior lobe is imperfectly developed. On the right side the inferior lobe is divided into two portions, the **medial** and **lateral lobules**, the inferior caval vein passing between them.

(d) The trachea divides at its posterior end into two portions, the **right** and **left bronchi,** one for each lung. Each bronchus is again divided into smaller portions, the **bronchial rami,** which penetrate the substance of the organ and redivide into smaller tubes. On the right side a small **eparterial bronchus** is given off from the right bronchus. It passes to the superior lobe of this side, and lies in front of the right pulmonary artery.

(e) The branches of the pulmonary artery and the pulmonary veins may be traced for a short distance on the medial portion, or **hilus,** of each organ.

(f) The vagus nerve passes to the dorsal side of the bronchus, giving off a number of branches, which accompany the bronchus to the lung.

The lungs may be removed, together with a portion of the trachea, care being taken to leave the vagus nerves in place. The lungs may then be examined to better advantage, and the surface also prepared for the next dissection.

5. The following structures may now be traced on the dorsal wall of the thorax:

(a) The oesophagus. It traverses the thorax in a median position, entering the diaphragm at the hiatus oesophageus.

(b) The vagus nerves. The right and left nerves pass backward along the lateral walls of the oesophagus, and are connected with one another through nerve plexuses lying on its dorsal and ventral surfaces. The left cord is that appearing at the posterior end of the oesophagus in a ventral position and passing to the ventral surface of the stomach. The right cord occupies a corresponding dorsal position and passes to the dorsal surface of the stomach (p. 185).

(c) The **thoracic aorta** (aorta thoracalis) passes backward on the ventral surface of the vertebral column, leaving the thorax through the hiatus aorticus, the latter being the aperture enclosed by the crura of the diaphragm. Its branches in the thorax are the **intercostal arteries** (aa. intercostales), beginning with the fourth, which are given off metamerically in the intercostal spaces, and pass laterad to the thoracic wall.

(d) The **thoracic portions** of the **sympathetic trunks** lie on the lateral surfaces of the bodies of the thoracic vertebrae. The posterior ganglia give origin to the **splanchnic nerve,** the latter passing backward into the abdominal cavity (p. 183).

(e) The **levatores costarum;** a series of small muscles arising from the transverse processes of the vertebrae and the heads of the ribs and inserted on the anterior margins of the next succeeding ribs. They assist the intercostals in respiration.

(f) The **intercostal nerves** (nn. intercostales) accompany the intercostal arteries to the lateral wall of the thorax.

(g) The **azygos vein** (v. azygos) is a small, asymmetrical, venous trunk lying to the right of the dorsal surface of the aorta. It receives the majority of the **intercostal veins,** which accompany the corresponding arteries and nerves, the tributaries extending backward to the first lumbar veins. It opens forward into the right superior caval. The more anterior intercostal veins are tributaries of the **right** and **left supreme intercostal veins** which open into the corresponding superior cavals.

6. The **diaphragm** (diaphragma) is a muscular and tendinous sheet forming the posterior wall of the thorax and separating the pleural cavities from the peritoneal cavity. It is somewhat dome-shaped in the relaxed condition, but in contraction it becomes flattened in such a way that the space occupied by the lungs is considerably increased, while the liver and related structures of the abdominal cavity are displaced backward.

As a muscle the diaphragm arises in three portions. The first, or **lumbar portion,** consists of two muscular and fibrous cords, the **crura,** the right much larger and stronger than the left, arising from the anterior spinous processes of the first three lumbar vertebrae. The second, or **costal portion,** arises from the internal surfaces of the posterior ribs. The third, or **sternal portion,** arises from the xiphoid process of the sternum. Its insertion is represented by its own tendinous central portion, or **centrum tendineum,** although the latter is virtually attached forward to the lungs and pericardium through the broad pulmonary ligament. The centrum tendineum is shaped somewhat like a trefoil, the fibres of the costal and sternal portions radiating outward from its margin.

The following may be made out on the posterior surface:

(a) The cut margins of the falciform, coronary, and left triangular ligaments.

(b) The **hiatus aorticus**, an aperture enclosed by the two crura and serving for the transmission of the aorta.

(c) The **hiatus oesophageus**, ventral to the foregoing, and serving for the passage of the oesophagus.

(d) The **foramen venae cavae**, situated slightly to the right and serving for the transmission of the vena cava inferior.

(e) The **superior phrenic arteries** (aa. phrenicae superiores) arise from the aorta at about the point of origin of the eleventh intercostals or from one of the latter, and enter the crura.

The inferior phrenic arteries are minute branches arising at the base of the coeliac.

(f) The **inferior phrenic veins** (vv. phrenicae inferiores), one on either side of the foramen venae cavae, at which point they enter the inferior cava.

The **superior phrenic veins** pass forward from the diaphragm, opening into the superior cavals.

XII. THE VERTEBRAL AND OCCIPITAL MUSCULATURE.

Dissection on the dorsal surface of the body from the occiput backward; also on the lateral and ventral surfaces of the neck.

The **serratus posterior muscle** lies on the dorsolateral surface of the thorax. It arises from the dorsal spinous ligament of the neck (ligamentum nuchae) and from the lumbodorsal fascia back to the last rib, and is inserted on the lateral surfaces of the eight posterior ribs.

The **splenius muscle** is a somewhat triangular sheet arising from the ligamentum nuchae and inserted on the supraoccipital and mastoid portions of the skull, extending also to the transverse process of the atlas.

These two muscles should be divided, the serratus posterior being removed from the surface.

1. The long muscles of the vertebral column.

Apart from the iliopsoas, psoas minor, and quadratus lumborum—muscles of appendicular insertion lying on the ventral surface of the vertebral column—the vertebral musculature com-

prises chiefly modified segmental muscles lying on the dorsal surface, for the most part in the area enclosed by the spinous and transverse processes of the vertebrae. They include the sacrospinalis, semispinalis, and intertransversarii. Their insertions are extended in part laterad to the ribs. In the cervical region they are represented by short muscles, separated for the most part from the thoracic and lumbar portions, and arising by accessory bundles from the anterior ribs, the corresponding thoracic, and the posterior cervical vertebrae. In the cervical region the muscles are easily separated from one another, but in the posterior part of the body it is necessary to dissect away the tough investment of lumbodorsal fascia which covers them.

(a) The **sacrospinalis.** Origin: Crest of the ilium and dorsal surface of the sacrum; mamillary processes of the six posterior lumbar vertebrae; investing lumbodorsal fascia.

This muscle is the largest and strongest muscle of the body. It extends forward over the surfaces of the ribs. Its medial border is separated from the middle line by a space of considerable width, in which the semispinalis and multifidus muscles are accommodated. In the lumbar region it is inserted in a continuous mass on the long transverse processes of the vertebrae and in the interspaces. In the thoracic region the muscle divides into two portions, namely, a thin lateral portion, the **iliocostalis** or **longissimus costarum,** and a thick medial portion, the **longissimus.** The latter receives in the posterior portion of the thorax strong accessory bundles from the semispinalis muscle on its medial side, the two muscles being inseparable at this point.

The iliocostalis is inserted laterally on the ribs as the **iliocostalis dorsi.** It receives medially from the ribs a number of accessory bundles, which are inserted forwards to the seventh cervical vertebra as the **iliocostalis cervicis.**

The longissimus is inserted by broad fleshy bands on the posterior margins of the ribs, medial to the accessory origins of the iliocostalis, this portion of the muscle forming the **longissimus dorsi.** Continuing to the neck it is inserted on the transverse processes of the three posterior cervical vertebrae, medial to the origin of the cervical portion of the

serratus anterior, but a number of accessory slips carry the insertion forward to the transverse process of the atlas. This portion is the **longissimus cervicis.** A separate band of fibres arising chiefly from the transverse processes of the second to fourth thoracic vertebrae joins the lateral, ventral portion of the splenius, and forms the **longissimus capitis.** It is inserted with the splenius on the mastoid portion of the skull.

(b) The **semispinalis** and **multifidus.** The band of muscle lying between the longissimus and the middle line, is composed of partly fused slips, arising for the most part by very long tendons from the mamillary and transverse processes, and inserted forwards on the spinous processes. It is divisible into two portions, which are superficially separated by a constricted area lying at the level of the last thoracic vertebra, this being also the point where the muscle is fused with the longissimus. The anterior portion, the **semispinalis dorsi,** is inserted by a series of fleshy slips on the spinous processes of more anterior thoracic vertebrae, but extends to the spinous process of the third or fourth cervical vertebra. The posterior portion, the **multifidus,** increases in size backward to the sacrum, where it is continuous with the abductor caudae posterior.

An almost separate muscle, covering the neck as a broad sheet immediately beneath the splenius and longissimus capitis, is the **semispinalis capitis.** It arises from the transverse processes of the five posterior cervical and the transverse processes of the first four thoracic vertebrae. It is lightly attached on a line from the transverse process of the atlas to the external occipital protuberance, but is inserted on the lateral surface of the latter. The more posterior and medial portion of the muscle is composed of separate slips arising in common with the longissimus capitis, two closely applied slips, however, at the free margin of the muscle, arising from the semispinalis dorsi and the longissimus. The principal, lateral portion is crossed by a tendinous inscription.

A second muscle, the **semispinalis cervicis**, is covered by the foregoing one. It arises from the articular processes of the posterior cervical and first thoracic vertebrae, and is inserted on the spinous processes of the cervical vertebrae, chiefly on that of the epistropheus.

(c) The **intertransversarii** are short muscles connecting the lateral portions of the vertebrae. They are distinguishable in part by their darker coloration. They increase in size backwards, being most conspicuous in the lumbar region, where they form thick muscular pads interposed between the mamillary and accessory processes. The last slip is attached to the crest of the ilium.

2. The following muscles constitute an occipital group, composed of short muscles arising from the atlas and axis and inserted on the atlas and the occipital portion of the skull.

(a) The **rectus capitis posterior superficialis.** Origin: Spinous process of the epistropheus. Insertion: External occipital protuberance.

(b) The **obliquus capitis superior.** Origin: Transverse process of the atlas. Insertion: Lateral surface of the occipital protuberance.

The foregoing muscles should be divided.

(c) The **rectus capitis posterior minor.** Origin: Posterior tubercle of the atlas. Insertion: External occipital protuberance.

(d) The **rectus capitis posterior major.** Origin: Spinous process of the epistropheus. Insertion: Laterally on the supraoccipital bone.

(e) The **obliquus capitis inferior.** Origin: Spinous process of the epistropheus. Insertion: Dorsal surface of the transverse process of the atlas.

(f) The **rectus capitis lateralis.** Origin: In common with the obliquus capitis superior, which covers it. Insertion: Base of the jugular process of the occipital.

3. Muscles of the lateral and ventral surfaces of the neck:

(a) The **scalenus anterior.** Origin: Transverse processes of the four posterior cervical vertebrae. Insertion: First rib.

(b) The **scalenus medius.** Origin: Tranverse process of the fifth cervical vertebra. Insertion: Lateral surfaces of the third to fifth ribs.

(c) The **scalenus posterior.** Origin: Transverse processes of the fourth to sixth cervical vertebrae. Insertion: First rib.

The medius is superficial; the anterior more or less separated from the posterior by the origin of the cervical portion of the serratus anterior.

(d) The **scalenus medialis.** Origin: The sternum at the point of attachment of the first rib. Insertion: Middle of the ventral portion of the first rib.

The foregoing muscles comprising the scalenus group, are destroyed by the removal of the ventral thoracic wall (p. 255).

(e) The **longus colli.** Origin: Bodies of the first six thoracic vertebrae. Insertion: The muscle passes forward on the ventral surface of the bodies of the vertebrae, giving off insertion fibres, and also receiving strands of origin. It terminates on the anterior tubercle of the atlas.

(f) The **longus capitis** is partly fused with the foregoing muscle, but its origin is in a more lateral position from the transverse processes of the first six cervical vertebrae. Insertion: Sphenooccipital synchondrosis.

(g) The **longus atlantis.** Origin: Lateral to the longus capitis, from the transverse processes of the third to sixth cervical vertebrae. Insertion: Transverse process of the atlas.

The longus capitis should be divided near its insertion.

(h) The **rectus capitis anterior.** Medial portion of the ventral surface of the transverse process of the atlas. Insertion: Basioccipital bone.

The **caudal musculature** in the rabbit comprises, in addition to the posterior extension of the cutaneus maximus, the following axial muscles:

(a) The **extensor caudae medialis.** It lies in the furrow between the spinous and articular processes of the posterior sacral and anterior caudal vertebrae, and is inserted on the transverse processes and dorsal surfaces of the caudal vertebrae.

(b) The **abductor caudae posterior** lies in the grove between the articular and transverse process and is inserted on succeeding vertebrae. It appears to continue the multifidus, but corresponds to the more medial portion of the longissimus.

(c) The **abductor caudae anterior.** Origin: Ischial spine. Insertion: Lateral surface of the sacrum and the transverse processes of the caudal vertebrae.

(d) The **flexor caudae.** Origin: Ventral surface of the sacrum and anterior caudal vertebrae. Insertion: Ventral surfaces of succeeding vertebrae.

These muscles are also known as sacro-coccygei, dorsalis, lateralis, and ventralis (a, b, d,) and coccygeus (c).

XIII. THE CENTRAL NERVOUS SYSTEM.

1. The spinal cord and nerve roots.

To expose the whole cord or a portion of it from the dorsal surface, the muscles should be removed on both sides of the vertebral arches and the latter broken away with the bone forceps. The following features may be made out according to the extent of exposure:

(a) The **spinal cord** (medulla spinalis) is a thick cylindrical white cord traversing the vertebral column in the vertebral canal. It is of almost uniform diameter, but exhibits two slight enlargements, one in the cervical, the other in the lumbar region. At about the middle of the sacrum it contracts to a slender filament, the **filum terminale,** which may be traced backward to the middle of the tail.

(b) The enclosing membranes or **meninges** of the cord are three in number, of which two are readily identified. Lining the internal surface of the bone is a thick fibrous investment, the **dura mater;** on the surface of the nervous matter a thin vascular membrane, the **pia mater.**
The dura and pia are connected by a loose web of connective tissue, the **arachnoidea.**

(c) The regional distribution of the nerve roots—eight cervical, twelve dorsal, seven lumbar, four sacral, and six caudal.

The cervical nerves are numbered from the vertebrae lying behind the intervertebral foramina from which they proceed, the remaining nerves from the vertebrae lying in front of the intervertebral foramina. The nerve transmitted by the intervertebral foramen between the seventh cervical and first thoracic vertebrae is described as the eighth cervical.

(d) The origin and primary divisions of the nerve roots may be worked out by removing carefully the lateral portions of the

arches of one or two vertebrae. The **posterior,** or **sensory root** (radix posterior), arises from the dorsolateral surface of cord, and expands immediately into a **spinal ganglion** (g. spinale). The more slender **anterior,** or **motor root** (radix anterior), arises by a number of filaments from the ventrolateral surface. The combined roots give origin to the posterior, anterior, and communicating rami. The **posterior ramus** is an inconspicuous branch (except in the first two cervicals) passing to the dorsal musculature and skin. The **anterior ramus** is the chief portion of the spinal nerve, the successive rami appearing as the components of the cervical and lumbosacral plexuses or as individual spinal nerves. The **ramus communicans** is a slender filament which passes ventrad to join the sympathetic trunk.

2. A small portion of the spinal cord may be excised and examined (preferably under water) for the following: (see p. 38, Fig. 19).

(a) The division of the cord into lateral halves by two median depressions, the **anterior median fissure** (fissura mediana anterior) and the **posterior median sulcus** (sulcus medianus posterior).

(b) The division of each half into three columns by shallow grooves, the **anterior** and **posterior lateral sulci.** The grooves are marked by the attachments of the anterior and posterior nerve roots. The three columns of each half of the cord are the **anterior, lateral,** and **posterior funiculi.**

(c) On the cut surface the **white substance** (substantia alba) is seen to form a peripheral investment enclosing the **grey substance** (substantia grisea) of the centre of the cord. The grey portion is somewhat H-shaped in section, each half being composed of anterior larger and posterior smaller masses, known in section as the **horns** of the grey matter, or as complete structures, the **anterior** and **posterior grey columns.** In the median plane, but nearer the dorsal than the ventral surface, is the minute **central canal** (canalis centralis), the cavity of the spinal cord.

3. The brain may be exposed by breaking away the supraorbital processes of the frontal bone and then removing the roof of the skull with bone forceps. In order to clear the brain and the roots of the cerebral nerves, it is necessary to remove the entire

lateral wall of the skull on both sides. The chief part of the operation consists in removing the temporal portion of the skull by successive steps, exposing first the flocculus of the cerebellum, a small stalked body which is almost completely enclosed by the dorsal portion of the petrosal. The entire petrotympanic bone is easily detached, and if removed en masse the paraflocculus and probably also the roots of the facial and acoustic nerves will be destroyed.

The arches of the first three or four cervical vertebrae should be removed, unless the anterior portion of the cord has been exposed in the previous dissection.

The spinal cord may be divided at about the level of the third vertebra. The brain should then be raised very gently from the ventral wall of the skull and the nerve roots divided with a small sharp knife or with fine scissors. This operation requires considerable care, since the nerves are strongly attached at their points of exit from the cranial wall, but very lightly attached to the brain, so that they are in danger of being torn away.

The anterior end of the brain may be freed by cutting across the small anterior expansions formed by the olfactory bulbs.

Portions of the dura mater removed with the brain may be cut away with scissors. It is adherent chiefly along two lines: one representing the longitudinal fissure between the cerebral hemispheres, the other the tentorium cerebelli. These parts contain the **superior sagittal** and **transverse venous** sinuses.

On the ventral surface of the brain as removed appear the basilar and internal carotid arteries and their branches. These vessels should be kept intact for examination at a later stage.

For the primary divisions of the brain see p. 71.

4. The superficial features may be identified as follows:

IN THE PROSENCEPHALON:

(a) The greatly enlarged **cerebral hemisphere** (hemisphaerium cerebri) forms with its fellow of the opposite side the largest portion of the brain. The two structures are separated by the **longitudinal cerebral fissure,** but are connected with one another by the commissures indicated below.

(b) The **olfactory bulb** (bulbus olfactorius) is a small expansion lying at the anterior end of each hemisphere. Its

cut end represents the point of origin of the **first cranial** or **olfactory nerve,** the various divisions of which may be found on the skull, where they may be traced into the perforations of the cribriform plate.

(c) The olfactory bulb is the anterior portion of the **olfactory lobe,** or **olfactory brain.** When traced backward on the ventral surface of the brain it is seen to be replaced by a white band of fibres, the latter forming the **olfactory tract** (tractus olfactorius). The tract ends posteriorly in an expanded portion of the brain, which, from its shape, is described as the **lobus piriformis.** The olfactory brain as thus defined is separated superficially from the remaining portions of the cerebral hemisphere by a longitudinal furrow, the **limbic fissure** (fissura limbica), which passes along its lateral margin. The anterior portion of the furrow, known as the **anterior rhinal fissure,** separates the olfactory tract laterally from the narrow anterior portion of the cerebral hemisphere. The corresponding posterior portion of the furrow, the **posterior rhinal fissure,** separates the lobus piriformis from the posterior enlarged portion of the hemisphere. The slight angle formed at the junction of the anterior and posterior rhinal fissures is the point of origin of a faint depression extending upward on the cerebral hemisphere. It represents a rudimentary, **lateral cerebral** (Sylvian) **fissure.**

(d) The **corpus callosum** is a broad white commissural band passing transversely from one hemisphere to the other. Its median portion may be exposed by pressing apart the medial margins of the hemispheres at the longitudinal cerebral fissure.

(e) The **pineal body** (corpus pineale) is a small, somewhat conical structure lying between the dorsal posterior tips of the cerebral hemispheres, and connected by a hollow stalk with the unpaired portion of the brain (the thalamencephalon) lying below it. The connection is concealed by a mass of pigmented vascular tissue, the beginning of the chorioid plexus of the third ventricle, and usually also by a small portion of the dura mater containing part of the sagittal venous sinus. The latter may be carefully detached.

By pressing apart the tips of the hemispheres and removing the pineal body, the dorsal surface of the thalamencephalon will be sufficiently exposed to make out the following features:

(f) The slit-like aperture appearing in the middle line after the removal of the pineal body represents the dorsal portion of the **third ventricle** (ventriculus tertius) (Fig. 84), the roof of which is formed anteriorly by the superficial portion of the **chorioid plexus** (plexus chorioideus), the latter entering the ventricle at this point.

(g) The lateral margins of the aperture are largely formed by two minute spindle-shaped masses, one on either side, the **habenulae.** Their posterior ends are united by a slender transverse band, the **habenular commissure** (commissura habenularum). The fibres of the commissure are faintly traceable forward, where they form a pair of thin whitish filaments (medullary striae).

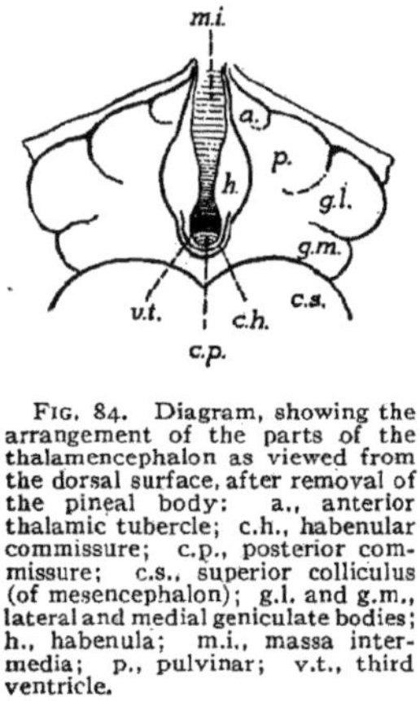

FIG. 84. Diagram, showing the arrangement of the parts of the thalamencephalon as viewed from the dorsal surface, after removal of the pineal body: a., anterior thalamic tubercle; c.h., habenular commissure; c.p., posterior commissure; c.s., superior colliculus (of mesencephalon); g.l. and g.m., lateral and medial geniculate bodies; h., habenula; m.i., massa intermedia; p., pulvinar; v.t., third ventricle.

(h) The **posterior commissure** (commissura posterior) crosses the posterior portion of the roof immediately behind and below the habenular commissure.

(i) The walls of the third ventricle are formed by the **thalami.** They are broadly connected by the **intermediate mass** (massa intermedia), or **middle commissure,** which may be seen from the dorsal surface crossing the cavity.

(j) On either side of the middle line, or of the habenulae, the dorsal portion of the thalamus forms a low, somewhat oval projection, the **pulvinar** or posterior tubercle.

The anterior tubercle of the thalamus is a faint elevation of very small dimensions lying in the angle enclosed between the pulvinar and the anterior portion of the aperture of the third ventricle.

(k) The parts of the **metathalamus** are distinguishable externally as two rounded projections of the lateral surfaces of the thalamencephalon. One of them, the **lateral geniculate body** (corpus geniculatum laterale), lies to the lateral side of the pulvinar, and is only separated from it by a faint depression of the surface. The **medial geniculate** body occupies a more posterior and medial position.

(l) The **optic tract** (tractus opticus) connects the geniculate bodies, especially the lateral one, with the ventral surface of the brain, ending in the optic chiasma.

(m) On the ventral surface, the **optic chiasma** (chiasma opticum), forms a conspicuous elevation, the posterior portion of which is traceable into the optic tracts, the anterior portion into the bases of the **second cranial,** or **optic nerves.**

(n) The **hypophysis,** or **pituitary body,** lies immediately behind the optic chiasma.

On account of its enclosure by the walls of the hypophyseal fossa, and also its slight attachment to the brain, the hypophysis is commonly detached in removing the brain from the skull, in which case a slit-like aperture, representing the ventral portion of the third ventricle is exposed.

(o) The **tuber cinereum** is a small elevation of grey matter appearing on the ventral surface after the removal of the hypophysis. It is the base of attachment of the **infundibulum,** the latter being the slender extension of the brain downward toward the hypophysis.

(p) The **mamillary body** (corpus mamillare) forms a conspicuous rounded elevation, lying at the posterior end of the tuber cincereum. The structure is single, but there is an indication of lateral lobes.

IN THE MESENCEPHALON:

(a) The dorsal surface is marked by four elevations, the **corpora quadrigemina.** The anterior pair, distinguished as the **colliculi superiores,** are much larger than the posterior pair, the **colliculi inferiores.**

(b) The ventral surface is occupied by a pair of divergent cords, the **cerebral peduncles** (pedunculi cerebri), separated by a faint median depression, the **interpeduncular fossa** (fossa interpeduncularis).

(c) The **third cranial,** or oculomotor nerve (n. oculomotorius), arises from the ventral surface of the cerebral peduncle.

IN THE RHOMBENCEPHALON:

(a) The **cerebellum** forms an arch over the posterior portion of the brain, and is supported by stout pillars from its sides. In addition to the ridges of its surface the whole structure is divided into several lobes, including the median portion or **vermis,** the lateral **hemispheres,** and the stalked bodies, or **paraflocculi,** arising from the hemispheres.

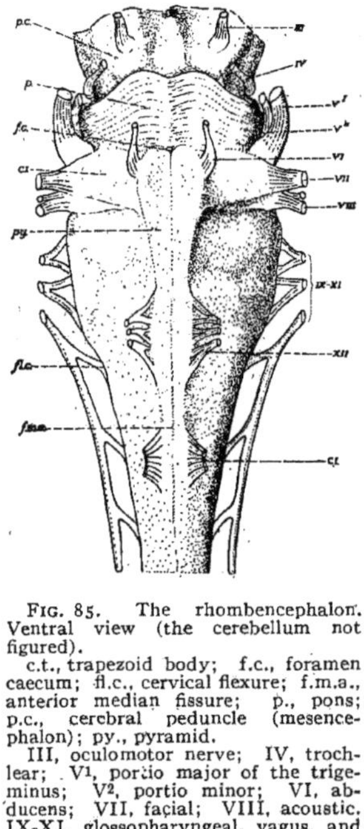

FIG. 85. The rhombencephalon. Ventral view (the cerebellum not figured).

c.t., trapezoid body; f.c., foramen caecum; fl.c., cervical flexure; f.m.a., anterior median fissure; p., pons; p.c., cerebral peduncle (mesencephalon); py., pyramid.

III, oculomotor nerve; IV, trochlear; V^1, portio major of the trigeminus; V^2, portio minor; VI, abducens; VII, facial; VIII, acoustic. IX-XI, glossopharyngeal, vagus, and spinal accessory group; XII, hypoglossal; CI, first cervical spinal.

(b) The **anterior medullary velum** (velum medullare anterius) is the thin membrane underlying the anterior portion of the cerebellum and connecting the latter with the inferior colliculi. It forms a small anterior portion of the roof of the fourth ventricle.

(c) The **fourth cranial,** or **trochlear nerve** (n. trochlearis), arises from the anterior medullary velum. The first portion of the nerve is usually to be found on the lateral surface of the cerebral peduncle.

(d) The **posterior medullary velum** (velum medullare posterius) underlies the posterior margin of the cerebellum, and extends backward over the triangular space enclosed by the walls of the fourth ventricle. The structure forms a chorioid plexus, similar in character to that of the third ventricle but much less extensive. It is commonly torn away in the preparation of the brain, in which case the interior of the fourth ventricle is exposed.

(e) On the ventral surface (Fig. 85), the **pons** forms a broad commissural band extending transversely across the brain and upward into the supports of the cerebellum. Its surface is divided into two parts by a median depression, the **sulcus basilaris,** occupied by the basilar artery.

(f) The anterior median fissure of the spinal cord ends at the posterior margin of the pons in a faint depression, the **foramen caecum.**

(g) The anterior funiculus of the cord is largely replaced forwards by the **pyramid** (pyramis), a narrow band extending forward on either side of the middle line to the posterior margin of the pons.

(h) The **trapezoid body** (corpus trapezoideum) lies in the angle formed by the lateral margin of the pyramid with the posterior border of the pons.

(i) The **fifth cranial,** or **trigeminal nerve** (n. trigeminus), arises by two roots, a larger sensory root, the **portio major,** and a smaller motor root, the **portio minor.** The two parts appear at the lateral border of the pons.

The portio major is the common trunk of the ophthalmic, maxillary and mandibular nerves, the portio minor joining the mandibular, so that the latter becomes a mixed nerve. The cut end of the portio major may be identified on the cranial wall and traced forward into the **semilunar ganglion,** the latter lying in a depression at the anterior ventral end of the petrosal bone.

(j) The **sixth cranial,** or **abducent nerve** (n. abducens), is a slender cord arising at the anterior end of the pyramid.

(k) The **seventh cranial,** or **facial nerve** (n. facialis), and the **eighth,** or **acoustic nerve** (n. acusticus), arise from the lateral margin of the trapezoid body.

The two nerves are closely associated, the former being slightly anterior in position. Its chief portion is the motor root. In addition the nerve receives a sensory filament, the **portio intermedia** or **intermediate nerve.**

(l) The **glossopharyngeus, vagus,** and **accessorius** arise by several roots arranged in a linear series along the lateral margin of the medulla. The trunk of the accessorius extends backward on the spinal cord, its roots, about ten in number, arising as far back as the fifth cervical spinal nerve.

(m) The **twelfth cranial,** or **hypoglossal nerve** (n. hypoglossus), arises by several roots from the ventral surface of the medulla

and at the lateral margin of the pyramid, its point of origin corresponding to that of the ventral root of a spinal nerve.

4. The arteries of the brain may be traced on its ventral surface as follows:

(a) The **basilar artery** (a. basilaris) is a median trunk formed on the ventral surface of the medulla oblongata by the union of the **vertebral arteries**, the latter here represented by their cut ends.

(b) The **inferior cerebellar artery** (a. cerebelli inferior) is the largest of several transverse vessels arising from the basilar and passing laterad to the cerebellum.

(c) The **posterior cerebral artery** (a. cerebri posterior) is formed on either side at the level of the anterior margin of the pons by the division of the basilar. It passes to the posterior portion of the cerebral hemisphere.

(d) The **superior cerebellar artery** (a. cerebelli superior) is a branch of the posterior cerebral, passing to the anterior portion of the cerebellum.

(e) The cut end of the **internal carotid artery** lies on either side of the tuber cinereum. It is connected backwards with the posterior cerebral.

(f) The **middle cerebral artery** (a. cerebri media) is given off from the internal carotid, passing to the middle portion of the hemispheres.

(g) The **anterior cerebral artery** (a. cerebri anterior), the continuation of the carotid, passes to the anterior portion of the ventral surface and the olfactory bulb. The anterior cerebral is connected with that of the other side, a complete anastomotic loop being formed around the hypothalamus by combined branches of the vertebral and carotid arteries. This is the **circle of Willis.**

5. By dividing the supports of the cerebellum on either side, the entire structure may be removed and the surface exposed, as in Fig. 86, for an examination of the structures of the dorsal surface of the rhombencephalon. The posterior medullary velum is removed with the cerebellum, but a small portion of the anterior medullary velum should remain in place.

(a) The **fourth ventricle** (ventriculus quartus) is the extensive space enclosed by the rhombencephalon. It is con-

nected forwards with the cerebral aqueduct and backwards with the central canal of the spinal cord. Its roof is formed principally by the anterior and posterior medullary vela, the latter being attached to the cerebellum close to one another, so that they underlie it.

(b) The **rhomboid fossa** (fossa rhomboidea) is the shallow depression enclosed by the thick lateral and anterior walls and floor of the ventricle. The middle line shows a narrow depression, the **posterior median fissure** (fissura mediana posterior), on either side of which the floor is raised into a low ridge, described as the **medial eminence** (eminentia medialis). The posterior end of the fossa forms with the enclosing wall the somewhat triangular figure described as the **calamus scriptorius**

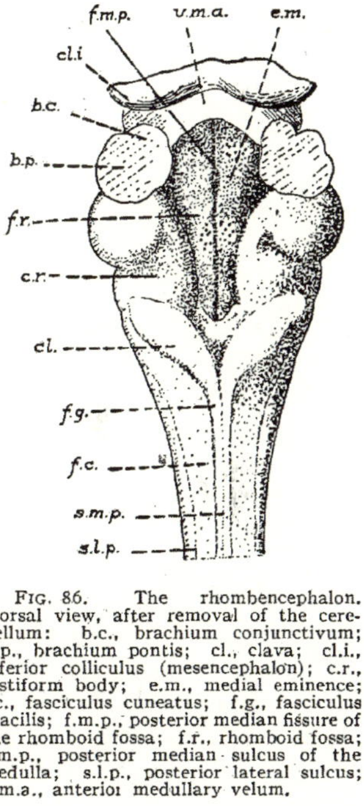

FIG. 86. The rhombencephalon. Dorsal view, after removal of the cerebellum: b.c., brachium conjunctivum; b.p., brachium pontis; cl., clava; cl.i., inferior colliculus (mesencephalon); c.r., restiform body; e.m., medial eminence; f.c., fasciculus cuneatus; f.g., fasciculus gracilis; f.m.p., posterior median fissure of the rhomboid fossa; f.r., rhomboid fossa; s.m.p., posterior median sulcus of the medulla; s.l.p., posterior lateral sulcus; v.m.a., anterioi medullary velum.

(c) The lateral supports of the cerebellum, now represented by their cut ends, are formed by fibre-bands passing into the cerebellum from adjacent ventral portions of the brain. In each a middle band, the **brachium pontis**, enters the pons; an anterior band, the **brachium conjunctivum**, enters the mesencephalon, while a posterior connection is established with the medulla through the restiform body.

(d) The lateral wall of the fossa is formed by a thick ridge of nervous matter, convex on its medial side, the **restiform body** (corpus restiforme). Its anterior portion bears a large spherical elevation overlying the origin of the acoustic nerve.

(e) The posterior funiculus, in passing forward from the cord, is divided into medial and lateral portions. The medial portion, the **fasciculus gracilis,** forms a narrow band terminating forwards in a club-shaped expansion, the **clava.** The lateral portion, the **fasciculus cuneatus,** passes into the restiform body.

6. The brain may be divided by a median vertical section, and one-half examined from the medial surface (Fig. 87). In addition to many of the features already made out on the surface the following may be noted:

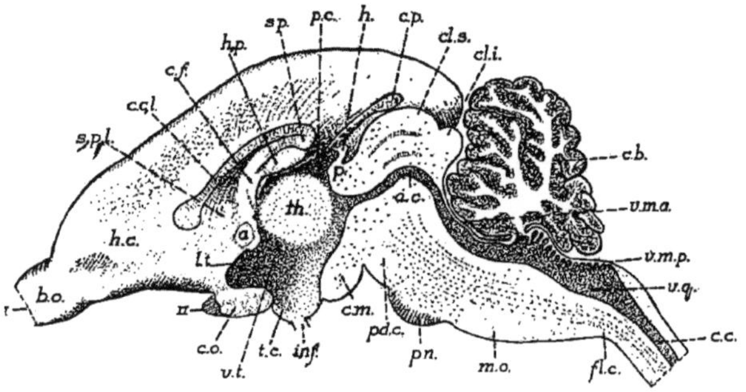

FIG. 87. The brain in median section: a., anterior commissure; a.c., cerebral aqueduct; b.o., olfactory bulb; cb., cerebellum; c.cl., corpus callosum; c.f., body of the fornix; cl.i., inferior colliculus; cl.s., superior colliculus; c.m., mamillary body; c.o., optic chiasma; c.p., pineal body; fl.c., cervical flexure; h., habenular commissure; h.c., cerebral hemisphere; hp., hippocampus; inf., infundibulum; l.t., laminal terminalis; m.o., medulla oblongata; p., posterior commissure; p.c., chorioid plexus of the third ventricle; pd.c., cerebral peduncle; pn., pons; sp., splenium; s.pl., septum pellucidum; t.c., tuber cinereum; th., thalamus, massa intermedia; v.m.a., anterior medullary velum; v.m.p., posterior medullary velum; v.q., fourth ventricle; v.t., third ventricle. I, olfactory nerve (origin); II, optic nerve.

(a) The deep but extremely narrow cavity formed by the third ventricle is the first space appearing in the brain from the anterior end, the first two (paired) ventricles lying laterally in the hemispheres. They are connected with the third ventricle by a narrow transverse canal, the **interventricular foramen** (foramen interventriculare).

(b) The anterior boundary of the third ventricle is formed ventrally by the narrow partition separating the two hemispheres, in the dorsal portion of which is the small **anterior**

commissure (commissura anterior). The ventral portion of the ventricle is projected toward the optic chiasma forming the **recessus opticus,** and into the infundibulum, forming the **recessus infundibuli.**

(c) The mesencephalon contains no ventricular expansion, its substance being perforated only by a narrow tube, the **cerebral aqueduct** (aquaeductus cerebri), which connects the third with the fourth ventricle.

(d) The corpus callosum is shown in section. Anteriorly it appears to end in a somewhat club-shaped expansion, but in reality is extended as a thin sheet of fibres downward toward the lamina terminalis. Posteriorly it bends downward, forming the **splenium,** the latter being attached to the body of the fornix, which lies below it.

The **fornix** consists of a pair of longitudinal fibre bands, fused for a short distance in the middle line to form the unpaired **body of the fornix** (corpus fornicis). They begin in the mamillary body, and passing upward as the **columns of the fornix** (columnae fornicis), meet in the body of the fornix, and afterwards diverge laterad as the **pillars of the fornix** (crura fornicis), ending in the hippocampus.

(e) Between the body of the fornix and the anterior portion of the corpus callosum is a thin area of the wall, the **septum pellucidum,** the lateral ventricles lying close together in this region.

7. The nervous matter covering the corpus callosum may be removed from one hemisphere by first marking out a triangular area on the dorsolateral surface; then scraping the material carefully away until the white surface of the corpus callosum is well exposed. By removing the corpus callosum the interior of the hemisphere may be examined.

(a) The **lateral ventricle** (ventriculus lateralis) is the extensive space enclosed by the hemisphere. It extends forward into the olfactory bulb and backward into the posterior free end of the hemisphere, passing a considerable distance behind the opening of the interventricular foramen.

(b) The excised portion of the hemisphere, forming the moderately thick roof and dorsolateral wall, consists largely of the peripheral grey cortex described as the **pallium.**

(c) The floor is formed by a greatly thickened mass of nervous matter, appearing from the interior of the ventricle in the form of two convex ridges. One of these, posterior and medial in position, is the **hippocampus.** The other is smaller, anterior and lateral in position, and is the **corpus striatum.** Between the two bodies the pigmented vascular tissue of the **chorioid plexus** of the lateral ventricle may be made out.

(d) On the medial wall, the thickened posterior portion forms the body of the fornix, immediately in front of which is the thinner portion of the wall, described above as the septum pellucidum.

8. The passage of the olfactory nerves to the ethmoturbinal surfaces may be traced by removing the nasal bones and working downward toward the cribriform plate,. or the remaining portion of the skull containing the nasal region still intact may be divided vertically for a more extended examination of the nasal fossae. The features to be observed are largely those described in connection with the skeleton (pp. 138, 151).

APPENDIX.

THE PRESERVATION OF MATERIAL.*

The method commonly used in the preparation of material for dissecting purposes consists in first embalming the body with suitable preserving fluids; afterwards filling the arteries with a colored injection mass, so that they are more easily traced. The objects served by embalming are: (1) preserving the body from decomposition for a sufficient length of time to complete the dissection; (2) keeping the body as nearly intact as possible; and (3) having the organs in good condition for study. The point last mentioned is an important one, since much depends on having the parts of the animal in such condition that they are easily and comfortably handled, and also easily observed. The desired results are accomplished, first, by introducing the preserving fluid through the bloodvessels, instead of by immersing the animal, as was formerly the practice; secondly, by using in the preserving fluid such materials as will leave the organs in a condition as near the natural one as possible and at the same time keep them moist and flexible throughout dissection

A suitable fluid for the purpose is that recommended by Keiller† for the preservation of human subjects. The formula is as follows:

Formalin	1.5	parts
Carbolic Acid	2.5	"
Glycerin	10.0	"
Water	86.0	"
	100.0	

---

*The methods here given apply only to the preservation of specimens for ordinary dissection, either singly, or in numbers for a laboratory course; with a few observations on the difficulties which are likely to be experienced. Especially in the matter of injections, the student who has acquired some knowledge of the vascular system will be able to make complete injections of the portal system and also satisfactory injections of the systemic veins, though the latter are somewhat more difficult on account of the presence of valves in the vessels. Finer vascular injections and injections of the lymphatic system according to the directions given in the anatomical textbooks may also be suggested.

†Keiller, W. "On the Preservation of Subjects, etc." (*American Journal of Anatomy*), Vol. II., 1902-3.

A convenient method of making up the fluid, especially when embalming the animals in numbers, is to prepare the mixture of formalin, carbolic acid and glycerin as a stock-solution, to be diluted for use by adding to each part of stock 6 parts of water. The amount required varies according to the size of the animal, the flow of the fluid in the vessels, the length of time during which the animal is left under the action of the fluid, and the height of the pressure column. Not less than 1500 cc. should be allowed for each specimen.

The apparatus needed for embalming includes a reservoir for the fluid, provided with an exit pipe to which a rubber tube may be attached; about 6 feet of rubber tubing to connect with the operating table; several three-way pieces to divide the stream in case several specimens are to be handled at the same time; selected rubber tubing of the size indicated below to attach the cannulae; clamps for the tubing; and, finally, glass or metal cannulae for insertion into the femoral artery.

Glass cannulae suitable for the purpose are readily made by heating ordinary glass tubing over the Bunsen flame and drawing it out to the desired thinness. The tubing used for the purpose should be of about 6 mm. outside diameter. The cannula when completed should be about 7 cm. long; and its narrow end should have a uniform diameter of 1.5-2 mm. for about 2 cm. at the tip. The tip should be touched lightly in the flame in order to round the margin by fusion, otherwise it might damage the wall of the vessel.

The rubber tubing used to connect the cannula with the main tube should be of the best quality of soft rubber, and should have an inside diameter of 4 mm., i.e., of proper size to slip on and off the cannula easily, but yet to retain its hold on the latter under moderate pressure.

The reservoir for holding the embalming fluid may be an aspirator or irrigator bottle, an enamel fountain, percolator or ordinary funnel. It may have a capacity of one or two quarts. The capacity, however, is immaterial, so long as the operator keeps the fluid replaced. The reservoir is suspended in such a way that it may be moved up and down within a distance of four feet above the top of the operating table.

At the time of beginning the embalming process the operator should have before him the reservoir, suspended at a height of about three feet, and a column of fluid, free from air-bubbles or foreign material to the tip of the cannula. This condition must be maintained throughout the operation. If at any time the pressure falls in the apparatus sufficiently to admit air, or allow coagulated blood to run back through the cannula, there is almost certain to be trouble, not only with the specimen under treatment, but also others which come after. The column of fluid is held back until the proper time by a clamp placed on the rubber tubing.

The animal is killed by administering ether or illuminating gas. It is placed on its back on the table, with the head away from the operator. The skin is first divided by a small incision on the inner side of the right thigh.* By inserting the fingers well down into the incision, the skin may be torn backward and toward the ventral middle line, and at the same time the superficial epigastric vessels will be carried with the subcutaneous tissue well out of the operator's way. Small portions of the inner surface of the thigh and of the abdominal wall will be exposed. The white cord representing the inguinal ligament lies in the bottom of the inguinal furrow. Appearing from beneath the ligament in this position, and passing to the surface of the thigh are the femoral nerve, artery, and vein, covered by an exceedingly thin layer of muscle belonging to the sartorius. The three structures may be separated from one another, and the muscle pulled away at the same time, by working lengthwise along the structures with the fine forceps. The artery must be thoroughly cleared for about 3 cm. from the inguinal ligament. Care must be exercised in this operation to avoid breaking its branches or the tributaries of the vein. The artery lies in front of the vein and is distinguishable by its smaller size, flattened or collapsed condition, and by its white coloration. The vein will be found greatly distended with blood. The nerve lies in front and partly on the lateral side of the artery.

When the femoral artery has been fully exposed, a ligature of coarse thread, previously moistened, may be passed around its base,

*The embalming may be done from the common carotid artery of the neck, a vessel much larger than the femoral artery and therefore easier of manipulation. This is not recommended, however, because of the damage done to various important structures of the cervical region.

close to the inguinal ligament. An ordinary single knot may be placed on the ligature, but must be left loose until the cannula is inserted. By grasping the bare edge of the artery at about 2 cm. from the ligament, the operator may make a V-shaped incision in the vessel with fine scissors. The tips of the scissors are directed toward the ligament. The incision must be clean-cut, and care must be taken not to cut more than half-way through the vessel. By taking up the little angular flap with the fine forceps, the cannula may be worked into the vessel and pushed well down into it beyond the inguinal ligament. The knot is then tightened by a gentle even pull on the ends of the thread. The knot should never be pulled very tight or doubled.

At the moment when the cannula is securely fastened into the vessel, the clamp is to be removed from the connecting tube and the fluid allowed to run in. At the beginning of the process a little care in arranging the animal will be amply rewarded by convenience in dissection. The hind limb on the side opposite the incision should be drawn backward. The front limbs should be drawn apart, so that the breast is well exposed, and held in this position by a thick cord, or, better, a stout flexible wire, passing around the back of the animal. The body should be turned slightly to the operator's left.

The animal is sufficiently embalmed in two hours. About eight animals may easily be kept on the table by one operator, provided he has at his disposal a sufficient number of cannulae, one for each specimen, since the first may be taken off the apparatus after the eighth has been put on. After some practice the number can be greatly increased.

Since small difficulties frequently occur in the process, especially in placing the cannulae and in keeping them clear of obstruction, a number of points may be mentioned which indicate to the operator just how the operation is succeeding. The entrance of the cannula into the artery, in the first place, is usually accompanied by a slight rise of blood into its tip. General muscle contractions in the recently killed animal are a safe indication of uniform flow of the fluid to these and also other parts of the body, and no clogging of the vessels need be feared. The fluid may usually be observed running in the cannula, and, of course, falling in the reservoir.

Finally, there are characteristic changes in the body. The abdomen becomes greatly distended, the subcutaneous tissue swollen, the eyes protrude, and there is usually more or less frothing at the nose. Leakage, either in the area of the incision or at the nose, is sometimes a sign of too much pressure. In the former case the leakage is frequently behind the cannula, and may be stopped by artery forceps. In the latter case there is no recourse but to confine the fluid to the nasal cavity by tying the nostrils.

After the embalming process the rubber tube is disconnected from the cannula, the latter being left carefully in place. The animal is then set aside, preferably for a few hours, but if wanted for use immediately the injection may be made after several minutes, usually with satisfactory results.

The injection mass may be made by mixing ordinary starch and water to the consistence of thin cream; then adding a finely-ground coloring material, such as vermilion or a very small quantity of carmine. There is some advantage in using a 5% formalin instead of water alone in making up this mass, the arteries having afterwards a brighter appearance, which is doubtless due partly to better preservation and partly to the fixing of the starch in the vessels. The mass must be thoroughly strained before use, in order to avoid the presence in it of particles which are too large to go through the cannula. The injection is made with a syringe, the latter being provided with a rubber tube of the same kind as that used in the embalming process. The mass is sent in by applying a gentle, even pressure, and it is sometimes advantageous to allow the injection to run backward and forward in the tube, each time applying a little more pressure. When the vessels have been filled in this way, the tube is clamped. By drawing on one cord of the ligature the knot is loosened sufficiently to withdraw the cannula, and by keeping a finger pressed on the end of the vessel, the knot may then be drawn tight without loss of injection.

It sometimes happens, despite ordinary precautions, that the cannula becomes clogged either with settled starch or with coagulated blood. In this case it may be easily removed, cleaned, and replaced. The same cannula should always be used.

Material prepared according to the directions given above will keep indefinitely, provided, however, that precautions are taken

to avoid contamination from the surface. These are especially necessary in view of the thick coating of hairs. It is a good plan, therefore, to sponge the animal with a preserving fluid which will penetrate the coat immediately, or if many specimens are being prepared, to immerse the whole animal for a moment. A suitable fluid for this purpose is formalin-alcohol, made by adding 2% of formalin to a mixture of equal parts of ordinary spirit and water. The alcohol ensures immediate penetration and assists the formalin in preservation. The fluid should be squeezed out of the coat so far as practicable. An excess is undesirable because the alcohol tends to withdraw fluid from the body if the animals are kept for some time before dissection, but more especially because the fluid is likely to get into the material during dissection where it has the effect of removing glycerin, so that the tissues become brittle and dry rapidly on exposure.

For the storage of material either before or during dissection no precaution is necessary except that of protecting the body from undue exposure to evaporation. The animals may be stored in a spirit tank if raised above the level of the fluid, or may be kept individually in special prepared boxes for convenience in the laboratory. A zinc-lined copper box with sliding top, of dimensions 6 x 6 x 20 inches, will be found to be adequate and of proper proportions for animals of average size.

INDEX

Spotted something that isn't quite right?

Let us know and we'll fix it for future reprints.
And as a thank you for your diligence, we'll
send you a free Sothis book of your choice.

publisher@sothis.press

Made in the USA
Coppell, TX
04 November 2026

11234121R00173

This page intentionally left blank

Except of course for that line above. And this one.
And whilst we have your attention, perhaps you'd
like to follow us on Instagram @sothis.press. Go on.

CONTENTS.

structure, together with experience of the laboratory practice upon which first-hand knowledge of gross anatomy depends. It is with reference to the latter group that the plan of nomenclature has been rigidly selected in accordance with the newer and more uniform terminology now happily established in the textbooks of human anatomy. The student will be warned against terminology as an end in itself, as against other short cuts to knowledge, and if properly directed will place the emphasis upon practice and initiative.

B. A. BENSLEY.

University of Toronto,
December 1st, 1920.

PREFACE TO THE THIRD EDITION.

The present edition of the "Anatomy of the Rabbit" appears in the same form as previously adopted for the second, but with some minor revisions and modifications. The necessity of reprinting the text after a comparatively brief interval of use, and the thorough revision in passing from the first to the second editions are together accountable for the fact that the number of corrections is not greater. Realizing that the ultimate value of a manual of dissection depends upon the combined experience of as great a variety as possible of instructors and students using it, the author has endeavoured at all stages of revision to incorporate such new ideas as have been received, and is appreciative of the interest taken in the matter by an increasing number of individuals.

The chief features of the text may be summarized as follows: The practical convenience of the rabbit is recognized as material for elementary anatomical study in the same way as in other fields of biological study and investigation. Though the principal design of the book is to direct the student in an orderly study of the structure of a mammal, points of physiological interest have been included so far as seemed advisable within the limits of a small manual of anatomical outlook. The setting of gross anatomy, in respect of microscopic anatomy, embryology, and the foundations of evolutionary development in general, has been carefully considered. While the points to which reference is made in the dissection are given a restricted description, there is no suggestion to the student of inadequate and superficial treatment, such as is common and perhaps necessary in manuals where the study of a number of types is presupposed as part of the course of instruction.

The book has been planned in part to serve the purpose of those zoological students who seek to obtain knowledge of a grade of organization sufficiently near that of the human body as a foundation for comparative studies, but more especially for premedical and medical students who by making a preliminary practical study of a convenient and easily obtained mammal may obtain thereby a knowledge of the foundations of human

PRACTICAL ANATOMY OF THE RABBIT

AN ELEMENTARY
LABORATORY TEXTBOOK IN
MAMMALIAN ANATOMY

By
B. A. BENSLEY, Ph.D.
Professor of Zoology in the University of Toronto

THIRD EDITION

TORONTO:
UNIVERSITY OF TORONTO PRESS
1921